Phase Equilibria With Supercritical Carbon Dioxide – Application to the Components of a Biocatalytic Process

Authored by Mercedes G. Montalbán and Gloria Víllora

Published in London, United Kingdom

Phase Equilibria With Supercritical Carbon Dioxide - Application to the Components of a Biocatalytic Process
http://dx.doi.org/10.5772/intechopen.100816
Authored by Mercedes G. Montalbán and Gloria Víllora

Contributors
Mercedes G. Montalbán, Gloria Víllora

First published in London, United Kingdom, 2022 by IntechOpen
IntechOpen is the global imprint of INTECHOPEN LIMITED, registered in England and Wales, registration number: 11086078, 5 Princes Gate Court, London, SW7 2QJ, United Kingdom

British Library Cataloguing-in-Publication Data
A catalogue record for this book is available from the British Library

Additional hard and PDF copies can be obtained from orders@intechopen.com

Phase Equilibria With Supercritical Carbon Dioxide - Application to the Components of a Biocatalytic Process
Authored by Mercedes G. Montalbán and Gloria Víllora
p. cm.
Print ISBN 978-1-80356-503-3
Online ISBN 978-1-80356-504-0
eBook (PDF) ISBN 978-1-80356-505-7

Meet the authors

Mercedes G. Montalbán graduated in Chemical Engineering and studied for her Ph.D. at the University of Murcia, Spain, focusing on the synthesis of silk fibroin nanoparticles using green solvents. During this period, she completed two predoctoral international stays at the University of Nottingham, UK with Prof. Peter Licence, and Tufts University, USA with Prof. David Kaplan. Thereafter, she was incorporated into the Chemical Engineering Department, University of Alicante, Spain by a postdoc contract, working in the field of biopolymeric nanocomposites. She became an assistant professor at the University of Murcia, Spain in 2021. She has co-authored more than thirty research papers and presented more than eighty communications at national and international conferences. She has also participated in eight research projects.

Gloria Víllora obtained her BSc in Industrial Chemistry from the University of Murcia, Spain. She obtained her Ph.D. from the same university and after completing postdoctoral research in chemical reaction engineering with Professor G. Froment at Ghent University, Belgium, she became a professor in 2010. Now, her core research is focused on developing new technologies within "green chemical engineering" and the production of nanomaterials using ionic liquids and supercritical CO2. She has directed about twenty R&D projects, maintained an intense collaboration with companies, and published more than 100 papers. She has been vice dean and dean of the Faculty of Chemistry and Coordinator of European Programmes, University of Murcia.

Contents

Preface

Currently, the aggregation states of matter are solid, liquid, gas, plasma, and Bose-Einstein condensate. A supercritical fluid is a quasi-state with intermediate properties between liquids and gases.

If a liquid is at rest in a closed container, an equilibrium inside the container will be established between the liquid and gaseous phases that will depend on the vapor pressure of the compound in question. If energy is applied to this system in the form of heat, the liquid will increase its temperature until it boils and the pressure inside the container will increase. Considering the phenomenon of thermal expansion of the liquid until it reaches its boiling temperature and the evaporation process, the density of the liquid will decrease while that of the gas will increase. If we continue with the process, we will reach a point where the densities of both phases become equal and it is not possible to differentiate the liquid phase from the gas phase. This point is well known as a critical point and is characterized by a critical pressure, temperature, and molar volume.

In the supercritical state, the density can be varied by modifying the conditions (since in the supercritical zone the density increases continuously), but the change takes place through a process of gradual transition from vapor to liquid, which contrasts with the sudden transition that occurs in condensation processes. This implies that once the critical point has been exceeded, no compression process is able to allow the liquid and gas phases to coexist. Therefore, it can then be said that, from the critical point, when the pressure increases, no liquefaction occurs, nor does evaporation occur when the temperature increases. Importantly, below and to the left of the supercritical region there are no definite boundaries and the transition from a single phase (gas or liquid) to supercritical fluid does not take place dramatically, but rather the change is gradual.

From an industrial point of view, supercritical fluids have a series of particular characteristics that make them very interesting for various applications. Their properties as solvents allow them to replace water and organic solvents in numerous processes, thus reducing their consumption. This is a great benefit for the environment, not only because of the water savings it implies but also because organic solvents are consumed in large quantities and are volatile and polluting substances. In particular, CO2 at high pressures, in liquid or supercritical state, is drawing great attention due to its worldwide expectations.

Another very interesting application of supercritical fluids is to use them as extraction agents since the products thus obtained are easily recoverable and the solvent can be easily reused by depressurization at atmospheric pressure. Due to these extraction properties, supercritical fluids are applied to obtain products such as decaffeinated tea or coffee, nicotine-free tobacco, pharmaceutical products, aromas, paints, antioxidants, polymers, and so on.

The separation with supercritical fluids requires two fundamental steps as occurs in the processes of separation, which use an auxiliary medium. Firstly, the desired extraction is carried out, and, secondly, the supercritical fluid is separated from the feed, regenerated, and recirculated. The separation of the supercritical fluid from the solute is normally an easy task that can be generally carried out by lowering the pressure and is usually accompanied by a lowering of the temperature. This can be also achieved by increasing the temperature at constant pressure. Extraction processes using supercritical fluids can be classified into three general types: single-stage, multistage, and continuous parallel-flow processes. These will be reviewed in the extended text.

An important application of supercritical fluids will be extensively revised. It is the use of supercritical fluids as an alternative to conventional organic solvents in biocatalytic processes, that is, in enzyme-catalyzed reactions. It was first investigated in 1985 and since then it has been a very fruitful area of research. The tunability of the solvent is perhaps the main characteristic of biocatalysis in supercritical fluids. In addition, the activity of enzymes in non-aqueous media depends on the properties of the solvent, thus supercritical fluids are attractive media to carry out biocatalytic reactions. Low viscosities and gas-like diffusivities increase the rate of mass transfer from reactants to enzyme active sites. Therefore, reactions whose rates are limited by mass transfer rather than intrinsic kinetics will be faster in supercritical fluids than in liquids. In addition, higher substrate concentrations can also increase the observed reaction rates and improve process performance.

A large number of experimental methods have been developed for the measurement of the phase equilibrium of fluids at high pressure. This is because high-pressure phase behaviour is often complex and difficult to predict and hence no single method is appropriate to study all different systems. As a general classification, the used methods can be divided into analytical or direct methods, and synthetic or indirect methods, depending on how the composition of the phases is measured. A section including a brief review of the techniques for measuring the high-pressure phase equilibria will be also discussed in the extended text.

A wide variety of high-pressure phase behaviours can be described. They can be classified into six types of pressure-temperature diagrams. Five of these diagrams can be generated from the van der Waals equations of state and will be widely described in the extended text.

At this point, the thermodynamic modelling of high-pressure fluid phase equilibria will be explained. For this, the most commonly used equations of state (van der Waals, Soave-Redlich-Kwong, Peng-Robinson, Patel-Teja) and the mixing rules needed to extend the equations of state to mixtures will be detailed.

Finally, the preceding concepts will be applied to the components involved in a biocatalytic process based on the transesterification reaction of 2-pentanol with a vinyl ester. This reaction has attracted growing interest from the pharmaceutical industry because (S)-2-pentanol can be obtained as a reaction product. This product is a key intermediate in the synthesis of drugs against Alzheimer's disease.

Mercedes G. Montalbán and Gloria Víllora
Faculty of Chemistry,
Chemical Engineering Department,
Regional Campus of International Excellence "Campus Mare Nostrum",
University of Murcia
Murcia, Spain

Chapter 1

Supercritical Fluids: Properties and Applications

Mercedes G. Montalbán and Gloria Víllora

Abstract

Currently, both humanity and the whole planet are living in a critical time, which leads us to look for more sustainable formulas to interact with the environment. One of the important changes in the design and operation of chemical processes is the search for environmentally friendly technologies. Many industrial processes are carried out under severe conditions or with reactants that involve the use of strong acids, toxic metal catalysts, organic solvents, and processes at high temperatures and/or pressures. Supercritical fluids (SCFs) and, among these, supercritical carbon dioxide ($scCO_2$), have been revealed as promising environmentally friendly solvents, energy-efficient, selective, and capable of reducing waste, constituting an alternative to conventional organic solvents. The use of SCF, such as solvents and reaction media, makes it possible to work in less severe and more environmentally friendly conditions, even considerably increasing the efficiency of the processes. This chapter provides a brief review of the most important properties of SCF, with special emphasis on $scCO_2$, as well as some of the most important applications.

Keywords: supercritical, critical properties, carbon dioxide, extraction, chemical reaction, nanoparticle, biocatalysis

1. Introduction

Currently, the aggregation states of the matter are solid, liquid, gas, plasma, and Bose-Einstein condensate. A supercritical fluid is a quasi-state with intermediate properties between liquids and gases.

The plasma is defined as an ionized gas, which is composed of electrons, protons, and free neutrons. It can be obtained by heating a gas or from a gas under strong magnetic fields. Although plasma is not naturally present on the Earth, the 99% of the matter of the universe is in the plasma state, with the Sun being our nearest example.

The Bose-Einstein condensate is an aggregation state, which is reached near to the zero absolute of temperature, i.e., −273°C. In this state, a macroscopic quantity of the material particles goes through the minimum energy level, which is well known as fundamental state. As a consequence of the Pauli exclusion principle, only the bosons can be in this state. Anderson et al. [1] synthesized this state at the laboratory in 1995, for the first time. For this, they cooled atoms at less than one millionth of a Kelvin above absolute zero of temperature.

In each state, matter presents differences in terms of the interaction forces between its molecules, which in the solid state are greater than in the liquid state and, in the latter, in turn, greater than in the gaseous state. This means that the molecules of a gas are less bound to each other than in the liquid and that their density is lower.

If a liquid is at rest in a closed container, inside the container an equilibrium will be established between the liquid and gaseous phases that will depend on the vapor pressure of the compound in question. If energy is applied to this system in the form of heat, the liquid will increase its temperature until it boils, and the pressure inside the contained will increase. Considering the phenomenon of thermal expansion of the liquid until it reaches its boiling temperature and the evaporation process, the density of the liquid will decrease while that of the gas will increase. If we continue with the process, we will reach a point where the densities of both phases become equal, and it is not possible to differentiate the liquid phase from the gas phase. This point is well known as critical point, and it is characterized by a critical pressure, temperature, and molar volume.

The critical temperature can be defined as the temperature above which the liquid-vapor meniscus cannot be formed when the pressure is raised at isothermal conditions. In the same way, the critical pressure can be defined as the pressure above which the liquid-vapor meniscus cannot be formed when the temperature is increased under isobaric conditions. Mathematically, the critical point is reached at a temperature and pressure at which the following is true:

$$\left(\frac{\delta P}{\delta V}\right)_T = 0 \tag{1}$$

$$\left(\frac{\delta P^2}{\delta V^2}\right)_T = 0 \tag{2}$$

As can be seen on phase diagram of **Figure 1**, the lines represent the different equilibria that can occur between the phases. Thus, the sublimation curve (for solid-gas equilibria), the melting curve (for solid-liquid equilibria), and the vaporization curve

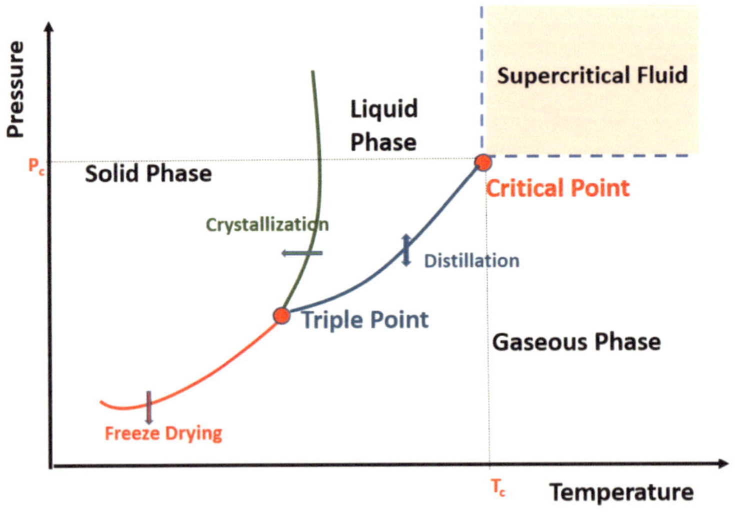

Figure 1.
Phase diagram including supercritical state.

(for liquid-gas equilibria) can be defined. The cutoff point for all of them is known as the triple point and represents the conditions in which the three phases coexist. The critical point is located at the end of the liquid-gas equilibrium line, and when these conditions, i.e., pressure and temperature, are exceeded, it is said that we have a supercritical fluid.

In the supercritical state, the density can be varied by modifying the conditions (since in the supercritical zone the density increases continuously), but the change takes place through a process of gradual transition from vapor to liquid, which contrasts with the sudden transition that occurs in condensation processes. This implies that once the critical point has been exceeded, no compression process is able to allow that liquid and gas phases coexist. Therefore, it can then be said that, from the critical point, when the pressure increases, no liquefaction occurs, nor does evaporation occur when the temperature increases. Importantly, it should be seen that below and to the left of the supercritical region there are not definite boundaries and the transition from a single phase, gas or liquid, to supercritical fluid does not take place dramatically, but rather the change is gradual.

In the supercritical state, matter is compressible and behaves like a gas, although it has a density like that of liquids (0.1–1.0g/cm^3). This property has led to supercritical fluids (SCFs) also being called "dense gases."

Regarding extracting operations with SCF, SCF can be named as "gases with density-dependent solvent power." In other words, the power as a solvent of a compound in supercritical state highly depends on its density and, therefore, on its pressure and temperature [2].

2. Properties of supercritical fluids

There are many substances commonly used as SCFs. **Table 1** collects some of them and their properties. CO_2 is marked in yellow.

From an industrial point of view, SCF have a series of characteristics that make them very interesting for various applications. Their properties as solvents allow that they can replace water and organic solvents in numerous processes, thus reducing their consumption. This is a great benefit for the environment, not only because of

FLUID	T_c (°C)	P_c (bar)	δ_c (g/cm^3)	Density at 400bar (g/cm^3)	Density of the liquid (g/cm^3)
Xe	16.6	58.4	1.10	2.30	3.08 (sat., 111.7°C)
CHF_3	25.9	46.9	0.52	—	1.51 (sat., −100°C)
CO_2	31.3	72.9	0.47	0.96	0.93 (63.4 bar, 25°C)
N_2O	36.5	72.5	0.45	0.94	0.91 (sat., 0 °C)
SF_6	45.5	37.1	0.74	1.61	1.91 (sat., −50°C)
CCl_2F_2	111.8	40.7	0.56	1.12	1.53 (sat., −45.6°C)
NH_3	132.5	112.5	0.24	0.40	0.60 (10.5 bar, 25°C)
Butane	152.0	37.5	0.23	0.50	0.58 (sat., 20°C)
Pentane	196.6	33.3	0.23	0.51	0.75 (1 bar, 25°C)

Table 1.
Physical properties (critical temperature, T_c, critical pressure, P_c, critical density, δ_c, of the most commonly used SCFs.

the water savings it implies, but also because organic solvents are consumed in large quantities and are volatile and polluting substances [3]. In particular, CO_2 at high pressures, in liquid or supercritical state, is drawing great attention due to its worldwide expectations [4].

In addition to being able to replace conventional solvents, SCFs have numerous advantages over them [5]:

- Its density can be modulated by adjusting pressure and temperature within the supercritical region, so the properties of the solvent can be varied just by modifying the operating conditions and without having to change the medium.
- They also have the high diffusivities inherent to gases (one or two orders of magnitude greater than those of liquids) and low viscosities (an order of magnitude lower than liquid solvents), which contribute to improving mass transfer processes.
- The supercritical processes consume less energy than those based on organic solvents, since they do not require working at high temperatures or the transitions between distillation and condensation.
- The use of SCF eliminates the environmental and safety drawbacks of organic solvents such as volatility, residue, flammability, and toxicity.
- The viscosity is much lower than that of liquids, which gives it very favorable hydrodynamic properties.
- They have high penetrability through porous solids and packed beds due to their low surface tension.

3. Supercritical carbon dioxide ($scCO_2$)

As it is widely known, CO_2 is a gas that is found in abundance in the Earth's atmosphere and is the product of the respiration of animals and plants. In addition, it is produced in large quantities in combustion processes and its accumulation is leading to the greenhouse effect and global warming. From the point of view of "green chemistry" or chemistry that respects the environment, the most interesting thing to use CO_2 in industrial processes would be to take it from the atmosphere or from the aforementioned combustion processes (e.g., in thermal power plants); nevertheless, in practice, almost all the commercial CO_2 is obtained as a by-product of the ammonia manufacturing process, after being properly purified [6].

The critical parameters of CO_2 are 31.3°C and 72.9 bar (see **Table 1**), which are easily achievable. This, together with its ease of transport and not having to work at very high pressures or temperatures (which would increase the danger of the process and the price of the equipment), makes it a relatively cheap and safe option, compared with the rest of SCFs. Among all its properties, those that have made it one of the most widely used SCFs in the world are its nontoxicity, its clean nature not generating residues, and that its critical temperature makes it suitable for working with compounds that do not resist high temperatures. For this reason, CO_2 can be used in the food and pharmaceutical industry. CO_2 has also a low reactivity (a very useful property if it is to be used as a solvent), it is not flammable (it makes it safe), and it is easy to recover and recycle. Its dielectric constant is ~1.5 for liquid and 1.1 ~ 1.5 for supercritical (depending on density).

In addition to the previous applications, $scCO_2$ is used in hydrogenations, reducing flammability and preventing fouling of the catalyst due to it can dissolve compounds that contribute to the formation of coke, in the production of polycarbonates as an alternative to the phosgene route, in the production of fine particles, in the production of certain polymers, in the foaming of thermoplastics, and as an extinguishing agent [7, 8].

Supercritical technology requires equipment with special specifications, which implies a high initial cost of the facilities, both for the main operating equipment and for the auxiliary control, prevention, and protection equipment. In addition, having a high pressure as a starting point could lead to problems when working with exothermic processes, since an increase in temperature could raise the pressure dangerously and a rigorous control of the process will be necessary. Some substances such as ammonia are toxic and flammable, and their use at high pressures and temperatures increases the risk of accidents and of their consequences. For these reasons, most governments do not authorize the creation of large industrial facilities that work with supercritical technology and that the vast majority of plants of this type are concentrated in two countries: China and India.

The low dielectric constant can be responsible of problems to dissolve polar substrates, although this can be avoided by adding a polar cosolvent such as ethanol, acetone, or water, which would increase the polarity of the solvent by hydrogen bonding. However, the presence of another compound may complicate reactions or have undesirable effects. Another option would be to complex the polar substrates with phenylboronic acid and their subsequent recovery by extraction for contact with an aqueous phase [9].

CO_2 is a Lewis acid, so it can react with strong bases such as amines, phosphines, or alkyl anions, which greatly complicates its use in reactions where these compounds are involved. Dense CO_2 produces low pH (around 2.85) if it is in the presence of water, since it leads to the formation of H_2CO_3. This can lead to problems in the field of biocatalysis because some enzymes are denatured under these conditions. This drawback can be turned into an advantage when you want to use H_2CO_3 as a reagent, since then you do not need a base to neutralize, but only to decompress. Finally, when $scCO_2$ is in the presence of noble metals, it can be hydrogenated producing CO, which is a poison for most catalysts.

4. Applications of supercritical fluids

4.1 Extraction with supercritical fluids

One of the most interesting applications of SCFs is to use them as extraction agents, because they can be used in separation operations at much lower temperatures than distillation, avoiding the degradation and decomposition of thermolabile compounds. The interest of SCF extraction processes is due to the possibility of regeneration, solvent versatility, energy savings with respect to other processes, selectivity, sensitivity of the results to operating conditions, improved solute volatility, solvent selectivity, favorable transport properties, and environmentally acceptable [5]. Furthermore, the products thus obtained are easily recoverable and the solvent can be easily reused by depressurization at atmospheric pressure.

By means of a simple depressurization, the SCF can be turned into a gaseous state, producing a phase separation and the precipitation of the extracted compound, which can be solid or liquid. In addition, in processes that require a fractional precipitation of several products, successive stages of decompression can be carried out in series.

Supercritical CO_2 ($scCO_2$) extraction is the most widespread industrial application of SCF. $scCO_2$ is applied to obtain products such as decaffeinated tea or coffee [10], nicotine-free tobacco [11], pharmaceutical products [12], aromas [13], paints [14], antioxidants [15], polymers [16], etc.

The separation with SCF requires two fundamental steps as occurs in the processes of separation, which use an auxiliary medium. Firstly, the desired extraction is carried out, and secondly, the SCF is separated from the feed, regenerated, and recirculated. The separation of the SCF from the solute is normally an easy task, which can be generally carried out by lowering the pressure, and usually accompanied by a lowering of the temperature. This can be also achieved by increasing the temperature at constant pressure.

Extraction processes using SCF can be classified into three general types. The simplest separation scheme, which is shown in **Figure 2**, consists of a single-stage process. It can be used to carry out the extraction of solids or liquids. For a successful extraction, very high separation factors are needed. Some examples of common industrial applications of single-stage processes are the extraction of oils, spices, and alkaloids from solids and the deoiling of lecithin [17].

Multistage processes can also be used for extraction. Their scheme is represented in **Figure 3**. This type of extraction is useful when the separation factors for one stage are not large enough. The main advantage of a multistage process is that the separation and/or extraction conditions can be modulated in order to improve the extraction. Examples of multistage processes include waste oil refining, edible oil fractionation, and deasphalting [17].

Finally, the third category of SCF extraction is the continuous countercurrent-flow operation, which is represented in **Figure 4**. This type of separation is similar to liquid-liquid extraction. Some examples of continuous countercurrent-flow operations include the enrichment of ethanol from dilute aqueous solutions, the separation of mono- and diglycerides, and the separation of fatty acids [17].

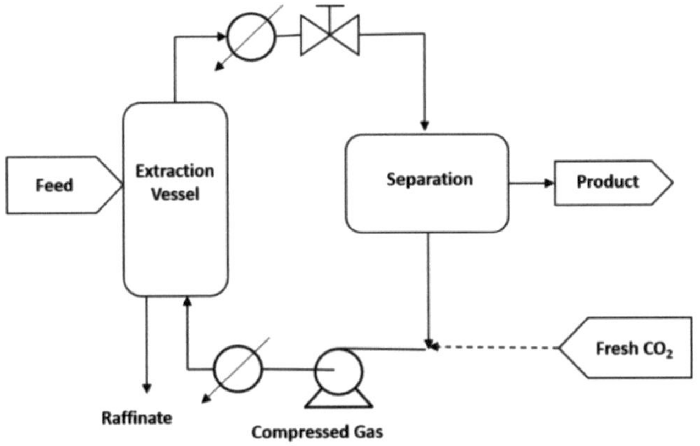

Figure 2.
Single-stage extraction process with SCF.

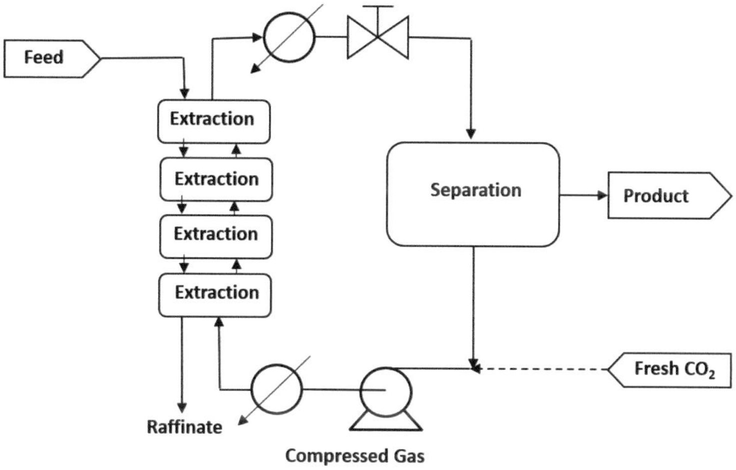

Figure 3.
Multistage extraction process using SCF.

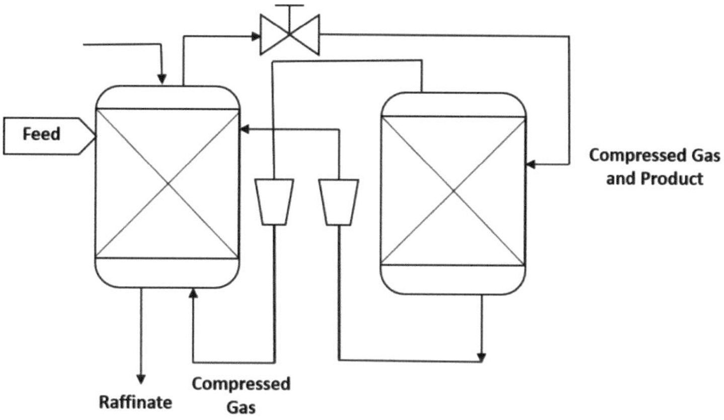

Figure 4.
Continuous countercurrent-flow extraction with SCF.

The decaffeination process is a typical example of supercritical extraction. This process provides products without caffeine or with a reduced caffeine content, such as decaffeinated coffee or decaffeinated tea. However, some processes leave a drink with low taste. Some methods use organic solvents, such as ethyl acetate, alcohol, chloroform, dichloromethane, or acetone. Other methods use water and $scCO_2$. $scCO_2$ has the advantages compared with the other methods that can retain most flavor compounds, is a nontoxic solvent, the process uses mild conditions (especially temperature), then, the energy consumption is generally lower than conventional processes, and CO_2 can be easily separated from the products, simply by decompression. In addition, $scCO_2$ has good dissolving capacity and good mass transfer performance. Because of that, $scCO_2$ decaffeination, despite being a relatively recent technology (was patented in Germany in 1964 and in the United States in 1974, and commercial plants went online in Germany in 1978 and in the United States in 1988), has expanded rapidly all over the world [18].

Decaffeination processes can serve to illustrate the different types of supercritical extraction from the point of view of their operating mode. One of these operating modes is by a batch process. In this case, the biomass and CO_2 are mixed in a fixed bed for a static time under established operating conditions (temperature and pressure). Once the determined time has elapsed, the fixed bed is depressurized and the extract and solid matrix are recovered. Generally, the batch process is carried out by a single stage due to the difficulty in handling the solids in pressure vessels by a continuous mode. **Figure 5** shows a typical process.

Another operating mode is by semicontinuous or semibatch process. As can be observed in the figure, in this type of process, the biomass is placed in a fixed bed and the CO_2 flows continuously through the bed under the established operating conditions. The extract is retrieved at the exit of the extraction vessel and the caffeine recollected. The most commonly process used in supercritical extraction is semicontinuous because pumps can be used to pressurize fluids without containing solid particles. As presented in the scheme of **Figure 6**, first the CO_2 must be change from gas to liquid phase to avoid cavitation of the pump, then the CO_2 is pressurized to operating pressure by a pump, heated to the objective temperature by a heat exchanger, and fed to the extractor containing the biomass. Finally, the depleted biomass is discharged for further drying. This type of process requires short residence times of $scCO_2$ in contact with the solid and allows the use of different particle sizes [18].

In continuous operation mode (**Figure 7**), the CO_2 and the solid biomass flow continuously through the extractor, so they must be mixed and pressurized by a pump to maintain the determined operating conditions of pressure and temperature. The extract is obtained at the exit of the extractor, and the decaffeinated solid material and the caffeine are separated. The continuous process needs pumps that pressurize fluids containing solid biomass particles. As can be seen in the figure, the sequence of stages that follow each other in this type of process is: CO_2 cooling; mixing of CO_2 with crushed biomass; pressurization of the CO_2 mixture with the biomass to

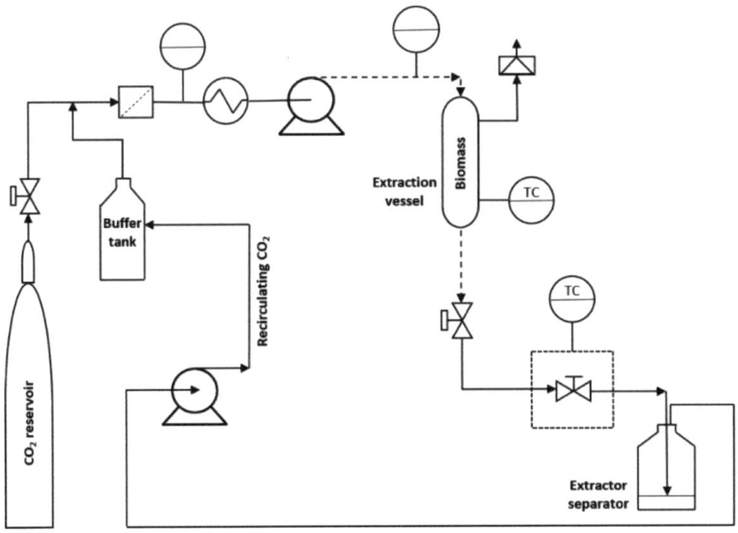

Figure 5.
Schematic flowchart of decaffeination by $scCO_2$ extraction in a batch process (dashed lines indicate a noncontinuous flow). Adapted from [18].

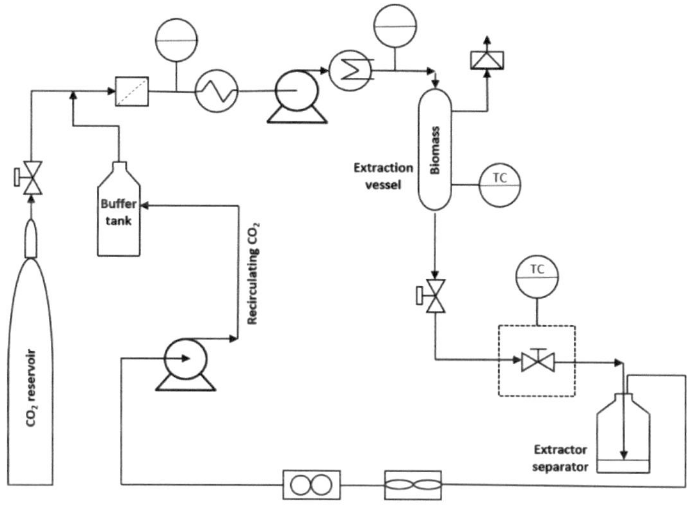

Figure 6.
Schematic flowchart of decaffeination by $scCO_2$ extraction in a semicontinuous process. Adapted from [18].

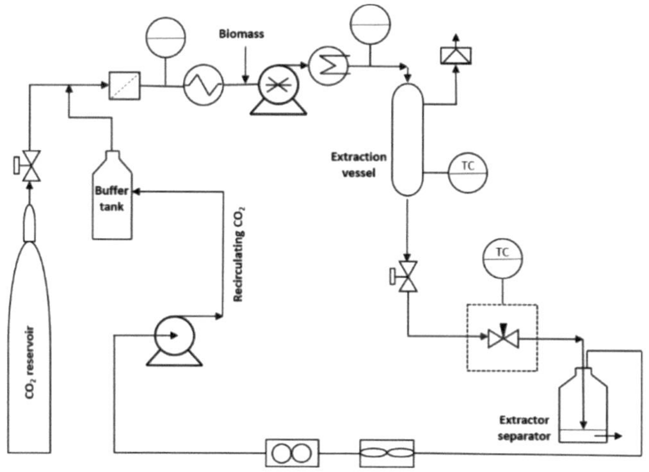

Figure 7.
Schematic flowchart of decaffeination by $scCO_2$ extraction in a continuous process. Adapted from [18].

the desired pressure using a high-pressure pump; heating of the mixture to the set temperature by means of a heat exchanger; pumping of the mixture to the extractor; the collection of the extract mixed with biomass at the exit of the extraction vessel in the separators. These systems are characterized by short residence times, high solvent-biomass ratios in the feed, very small sizes of solid particles, and good process control [18].

For large extraction of caffeine processes, depressurization to atmospheric pressure makes solvent reuse more expensive, since compression processes are energy expensive. Therefore, a liquid-liquid extraction with water is preferred, so it is not necessary to reduce the pressure so much to extract the desired product. Furthermore, $scCO_2$ with recirculated water has better extraction properties than dry CO_2 [19]. **Figure 8**, adapted from [20], shows a scheme of the Kraft General Foods' patented process for decaffeinated green coffee beans. Saturation with water has been shown to improve caffeine extraction rates, and increased temperature and pressure improve caffeine partitioning in the supercritical phase. The extraction of $scCO_2$ from many compounds from natural substrates (plants, algae, materials of animal origin, etc.) has been investigated. Examples include extracting α acids from hops, flavors, spices, and fragrances that have been extracted and include lilac, essential oils, black pepper, nutmeg, vanilla, basil, ginger, paprika, rosemary, chamomile, and ground chili peppers.

Large-scale industrial applications require specific, single-purpose plant designs. However, multipurpose $scCO_2$ extraction plants can be designed to be used for obtaining different extracts of high added value through processes that operate at smaller scales. A common element of these plants that is important for the economy of the process is the system used to separate and recirculate CO_2 to the extraction unit.

The recirculation cycle may be carried out with a liquid pump or with a compressor. The recirculation cycle with a liquid pump consists of condensing the carbon dioxide at the output of the process and recirculating it using a pump for liquids. To condense carbon dioxide, the pressure must be reduced below the critical pressure and the carbon dioxide must be cooled. The pressure and temperature conditions used for condensation and recirculation must be optimized in each specific case, but pressures in the range of 40–60 bar are recommended, since the corresponding condensation temperatures are close to the ambient temperature, which allows the use of cheap refrigerants such as cooling water. When the process requires mixing an organic solvent with CO_2, it is necessary to take into account the separation of the same and the purification of CO_2. It is possible to achieve separation of the organic solvent by partial condensation of the supercritical effluent prior to CO_2 condensation, but if the CO_2 purity achieved is not high enough, more complex and expensive processes such as adsorption or cryogenic distillation may be necessary. The recirculation cycle with

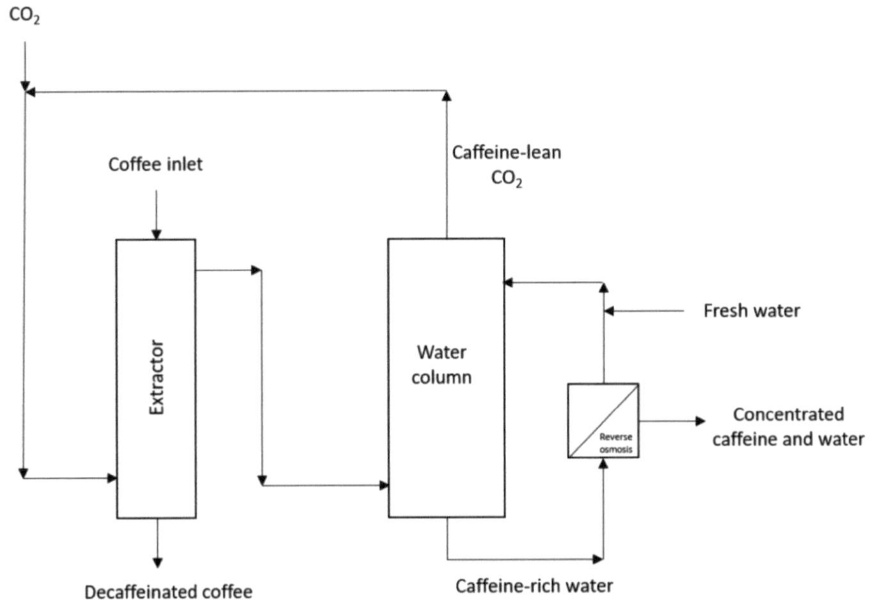

Figure 8.
Scheme of semicontinuous caffeine extraction process using $scCO_2$. Adapted from [20].

a compressor consists of recompressing and recirculating the effluent directly in a gaseous or supercritical state. If CO_2 must be purified before recycling it, it is necessary to include additional stages. The choice of using a compressor or liquid pump for the recirculation cycle depends on economic and technical reasons and will depend on the type of process. In general, the liquid pump is preferred for small installations for reasons of economy and simplicity of use [21].

Another very important factor to consider when designing an extraction process with $scCO_2$ is the pressure of the extractor. The most common values are in the range of 5–30 MPa. As for the container, fast-closing designs, different sizes, and (1 L-1 m^3) and materials (metallic and polymeric) are available on the market.

Although the investment costs of a $scCO_2$ extraction plant can be high, the operating costs are usually relatively low, due to the reduction in energy consumption related to the use of temperatures close to the environment [22].

4.2 Formation of particles by supercritical technology

After more than three decades using SCF for particle formation, several techniques have been developed. Technologies currently used differ in the basic principles of operation, but they are all based on the mechanisms of nucleation, particle growth, and precipitation. In this section, we will give a brief overview of the different processes of particle formation using SCF.

Table 2 summarizes the characteristics of the different particle formation technologies using SCF.

4.2.1 Rapid expansion of supercritical solutions (RESS)

This technique is used when the material to be used has certain solubility in the SCF. In an RESS process, the SCF acts as a carrier of the solute, and this solution expands adiabatically leading to a rapid decrease in temperature and pressure and the subsequent generation of small particles after being sprayed through a nozzle. This process is based on the difference in solubility of the polymer in the fluid at high and low pressure. The schematic principle that governs this expansion is shown in **Figure 9**. As is known, the depressurization of a gas leads to a reduction in temperature. Depending on the temperature of the process, and the glass transition or

Process	Role of supercritical fluid	Role of organic solvent	Mode of phase separation
RESS	Solvent	not needed or cosolvent	pressure/temperature induced
GAS	Antisolvent	solvent	solvent-induced
SAS	Antisolvent	solvent	solvent-induced
SEDS	Antisolvent	solvent/ antisolvent	solvent-induced
PGSS	dispersing agent solute	not needed	pressure/temperature induced
PGSS-drying	solute/propellant	not needed	pressure/temperature induced
SAA	Propellant	solvent or not needed	pressure/temperature induced

Table 2.
Summary of particle formation technologies using SCF. From reference [21] with permission of John Wiley & Sons.

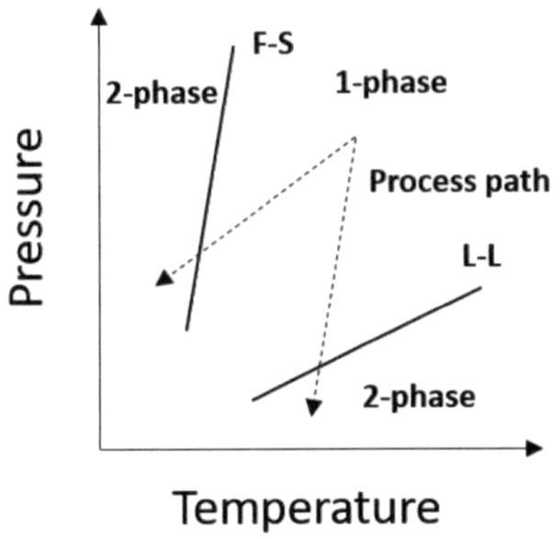

Figure 9.
Schematic representation of the operational principle of the RESS process. Adapted from reference [23].

fusion temperature of the polymer and the degree to which these transitions may have been reduced and the path followed by the homogeneous region of a phase, particle formation may come from crossing the fluid-solid barrier (F-S), or the system may have crossed the liquid-liquid barrier (L-L), followed by solidification [23]. Then, when designing this process, the solubility of the material plays a crucial role in the formation and processing of particles, since most pharmaceutical substances, such as polymers, drugs, and proteins of high molecular weight, are polar in nature. In some cases, small amounts of organic solvents are added to improve the affinity of the polar molecules of the drugs.

RESS is the simplest and most effective method of SCF technology, but its application is limited due to its relatively high cost and, when $scCO_2$ is used, the low solubility of most solutes in non-polar $scCO_2$. To solve this problem, progress in the RESS process has been made to overcome these limitations. One of them is the RESS process in an aqueous solution containing a surfactant or other reducing agents, known as the process of rapid expansion of a supercritical solution in a liquid solvent (RESOLV), where the SCF expands into a liquid medium. This modified process inhibits the agglomeration of particles in the expansion jet. **Figure 10** shows a scheme of the RESS process.

4.2.2 GAS antisolvent process (GAS)

This process is suitable for recrystallizing solids that are not soluble in SCF. This technique has been widely used with polymeric materials because most of them are not soluble in SCF or gases. In this process, the polymer is first dissolved in organic solvent and a gas is used as the antisolvent. The gas is injected into the airtight cell containing the solution, and precipitation of the particles takes place as the concentration of the gas in the solution increases with pressure. In this technique, it is not necessary for the anti-insolvent gas to be in supercritical conditions. The controlling principle of the process is solvent-induced phase separation, which is illustrated in **Figure 11**. With the introduction of the anti-dissolver, the F-S and L-L phase limits are displaced at higher temperatures or pressures, respectively. As a result, the system that was initially in a single homogeneous phase is now in a region of two phases, this separation of phases leads to the formation of particles. A significant difference between the RESS and GAS processes is that while in the first one it is treated with a binary system, material + SCF, in the second, we have a ternary system composed of

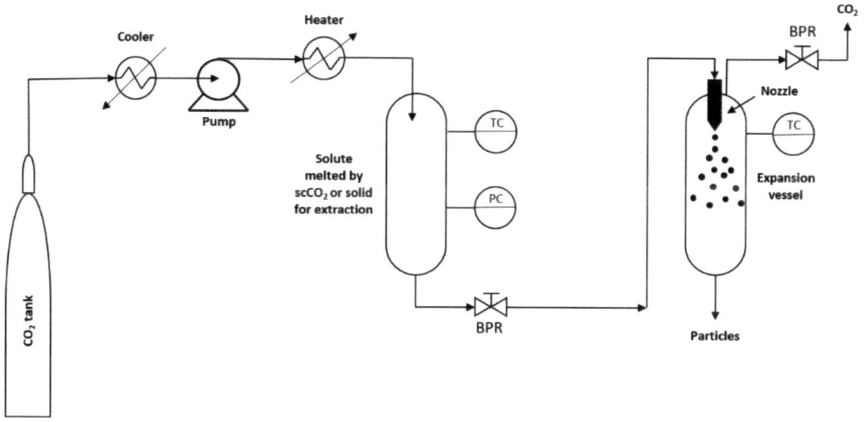

Figure 10.
Scheme of an RESS process.

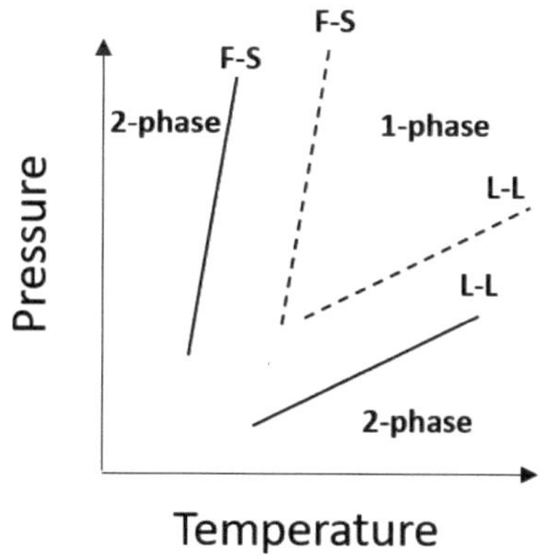

Figure 11.
Schematic representation of the operational principle of the GAS process. Adapted from reference [23].

the material + the organic solvent + the antisolvent (gas or SCF) [23]. **Figure 12** shows a scheme of the GAS process.

4.2.3 Supercritical antisolvent process (SAS)

In this technique, as in the GAS process, the SCF acts as an antisolvent for a solution containing a material, but in this case the mechanism is different. The compound to be micronized is dissolved in a liquid, generally an organic solvent, and the solution is injected into a high-pressure vessel where the supercritical antisolvent is, reaching intimate contact between the two media quickly. By mixing these two currents, the solvent is extracted by the SCF phase, and the SFC is simultaneously diffused into the liquid solution. As SFC is a poor solvent for the compound, in this mixture it acts as an antisolvent. Consequently, the mixture is oversaturated, resulting in rapid nucleation and growth. This process can operate in batch or semicontinuous mode, but a continuous operation configuration can also be adopted, which is important for industrial-scale production. The process is governed by solvent-induced phase separation, as described in **Figure 11**. **Figure 13** shows a scheme of the SAS process.

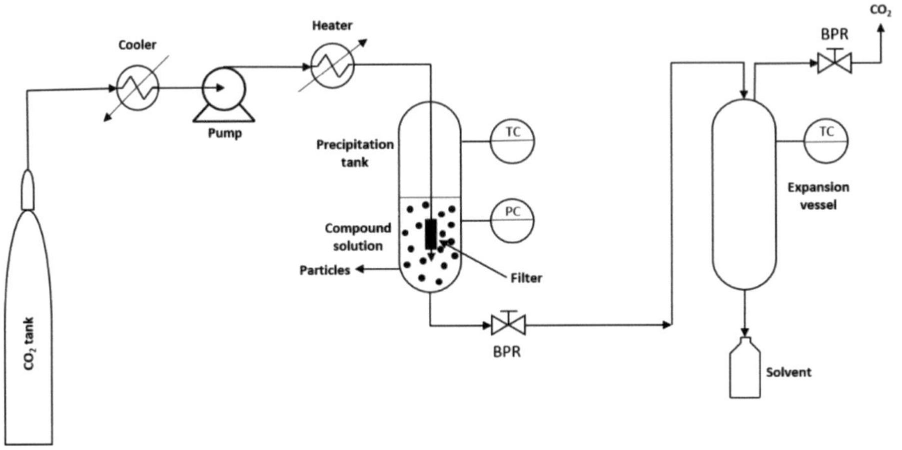

Figure 12.
Scheme of a GAS process.

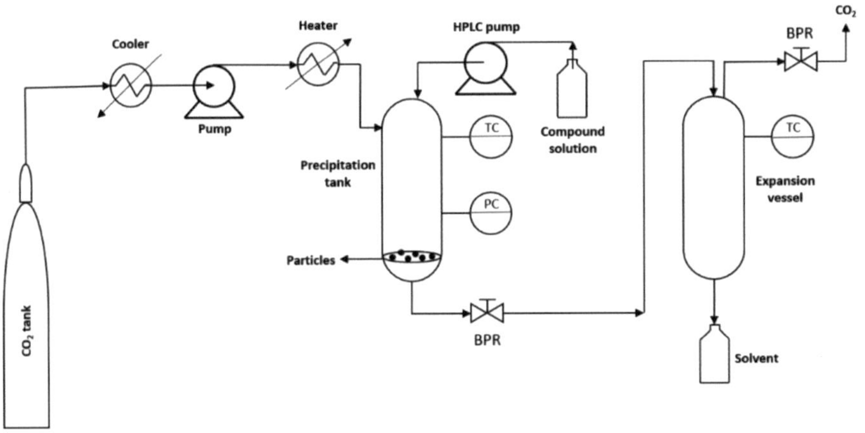

Figure 13.
Scheme of an SAS process.

In $scCO_2$ processes, the kinetics of particle formation using a supercritical antisolvent technique is fast compared with that of liquid antisolvent techniques, due to the good transport properties of $scCO_2$. Furthermore, it is possible to accelerate the rate of the process operating under conditions above the critical point of the CO_2-solvent mixture, since in these conditions the solvent and CO_2 are completely miscible and the diffusional limitations in the mixture disappear, establishing favorable conditions to produce ultrafine particles with a narrow distribution of particle size. Even by adjusting the operating conditions (pressure, temperature, and initial concentration of solutes), the morphology of the particles can be modified [21].

4.2.4 Solution-enhanced dispersion by supercritical fluids (SEDS)

The SEDS process was developed by York and Hanna of Bradford University in the year 1996 [24] to improve the efficiency of the traditional SAS process. The SEDS process usually operates at a lesser time with increased mass transfer rates.

The main objective of the SEDS process is to produce uniform-sized fine particles in a single-phase equilibrium, while removing the organic solvent to obtain them in a dried form. In SEDS, the liquid solution and SCF are sprayed together using a coaxial injector. These injectors can be found in two- and three-channel versions, which are used for precipitation of one or two components respectively. In this process, the SCF is used as an antisolvent and dispersing agent of the medium. High-speed contact between the liquid and the SCF generates a finely dispersed mixture and rapid precipitation of particles. As in the GAS and SAS processes, the basic principle here is also solvent-induced phase separation. Particles are formed due to a change in composition in phases (**Figure 11**). The design of the nozzle, as well as its specifications, particularly internal diameters, is decisive in the particle size distribution. In addition, operating conditions such as the flow rates of compound dissolution and SFC and critical parameters must be optimized to control both particle size and morphology [25]. **Figure 14** shows a scheme of the SEDS process.

4.2.5 Particles from gas-saturated solutions (PGSS)

This process is suitable to generate particles of compounds that absorb SCF in high concentrations. The PGSS process is quite similar to the RESS process. In the PGSS process, an SCF is dissolved in a molten polymer or in a liquid suspension and the mixture, which is at high pressure, is quickly depressurized through an injector leading to the formation of particles. This process has generally been used for the coating of compounds on polymer matrices. The phenomena that control this process are: the rapid changes in pressure and temperature and the solvent-induced phase separation (**Figure 15**).

The PGSS process has the advantage with respect to RESS that it is not necessary for the substance to be soluble in SCF, the process is simple, and low cost and technique avoid the low solubility in SCF of many molecules of pharmaceutical interest such as proteins and peptides that would be too difficult to treat with RESS. Although the applications currently developed use polymeric materials, it could be used to obtain powdered polymers or to apply coatings. **Figure 16** shows a scheme of the PGSS process.

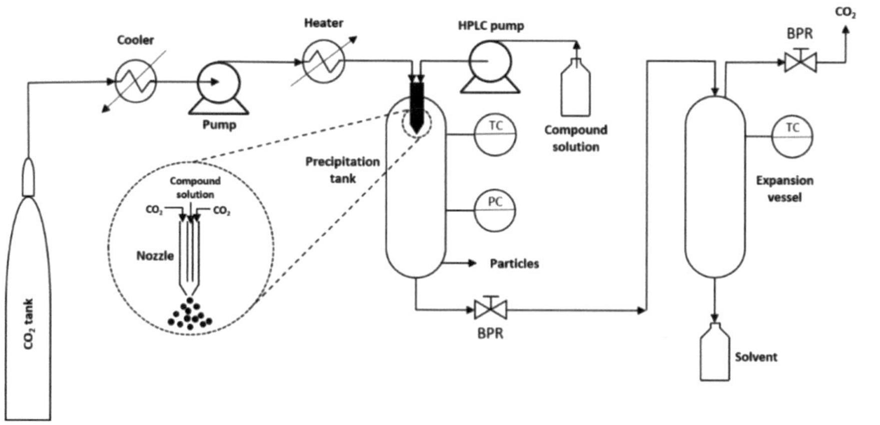

Figure 14.
Scheme of an SEDS process.

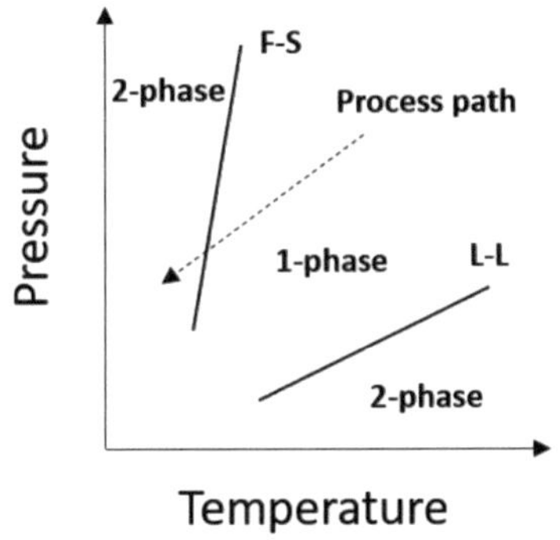

Figure 15.
Schematic representation of the operational principle of the PGSS process. Adapted from reference [23].

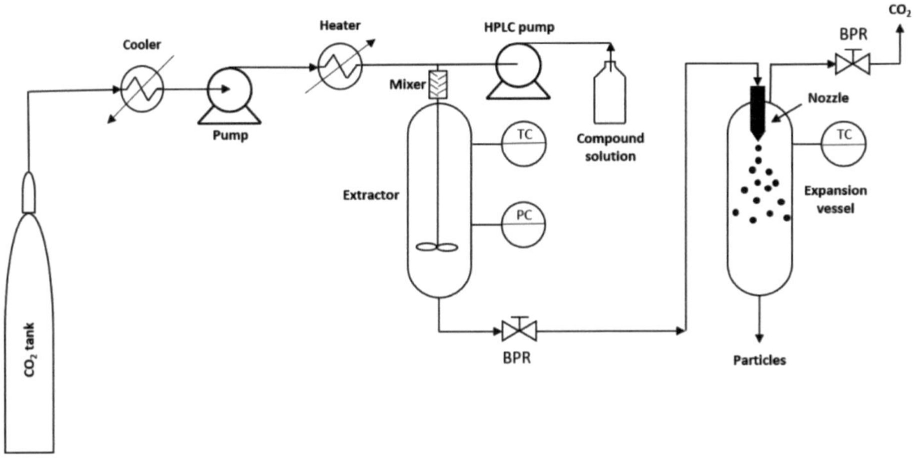

Figure 16.
Scheme of a PGSS process.

4.2.6 Particles from gas-saturated solutions drying (PGSS drying)

PGSS drying technique is a modification of the standard PGSS process, designed specifically to treat aqueous solutions (**Figure 17**). The process consists of pressurizing and preheating the aqueous solution and the CO_2 to the conditions prior to expansion. They are then vigorously mixed in a static mixer, causing partial water extraction and the saturation of the aqueous solution with CO_2. The resulting mixture is sprayed through a nozzle in a thermally insulated spray tower. In this tower, CO_2 vaporization and the expansion of gas bubbles due to depressurization occur. The water then evaporates into the CO_2 expansion vessel, thus producing particles. These are collected at the bottom of the tower, and CO_2 with evaporated water leaves the tower through its top. A cyclonic separator can be used to recover fine dust trapped in the gas effluent [26].

4.2.7 Supercritical fluid-assisted atomization (SAA)

The SAS technique is based on the solubilization of controlled amounts of $scCO_2$ in liquid solutions containing a solid solute and the atomization of the resulting solution

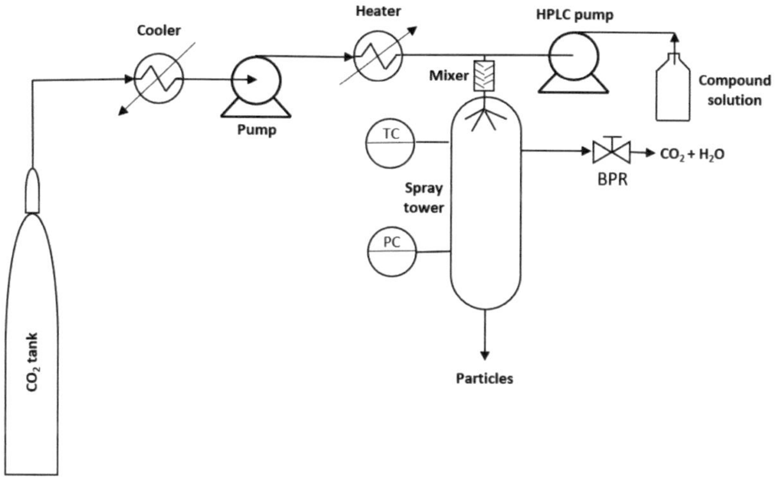

Figure 17.
Scheme of a PGSS drying process.

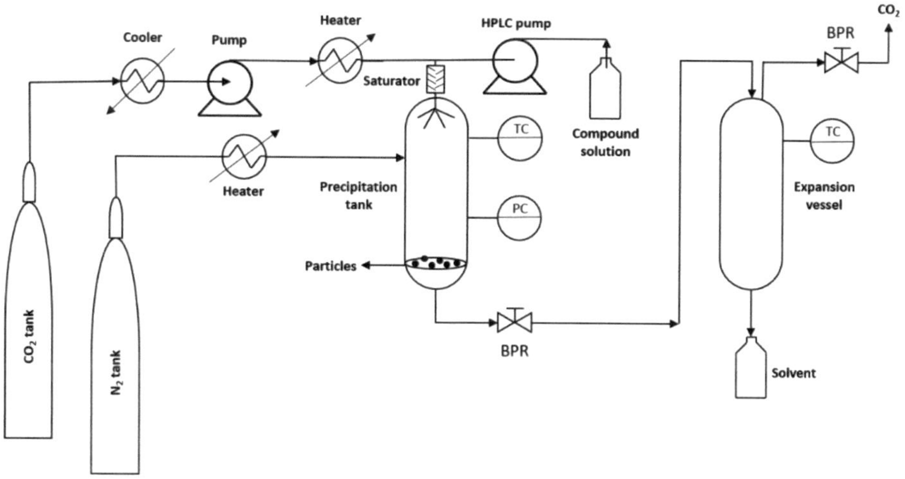

Figure 18.
Scheme of an SAA process.

through a nozzle. The experimental device is shown in the **Figure 18**. It consists of three lines that feed $scCO_2$, the liquid solution and an inert gas, such as N_2, and three main process vessels: saturator, precipitator, and condenser. The CO_2 is supplied in liquid phase to a high-pressure pump and sent to a heated bath. Next, it is taken to a contactor in which the CO_2 is solubilized in the liquid solution. The liquid solution is pressurized in a high-pressure pump, heated, and sent to the saturator. Subsequently, a spray is produced that forms the droplets in the precipitator. The inert gas is heated in a heat exchanger and sent to the precipitation tank in order to promote the evaporation of the liquid solvent. Particles are collected in a frit located at the bottom of the precipitator, and gases are discharged into a refrigerated condenser to obtain the liquid solvent. This technique has been used with compounds such as precursors of superconductive materials and catalysts and with pharmaceutical and ceramic compounds [27].

4.3 Reactions in supercritical fluids

The application of SFC to the extraction of thermally labile compounds is the most widespread. However, recently, they have also been widely used as reaction media. The properties of SFC can be harnessed to be used as reaction media in different processes. A supercritical fluid can be used to reduce temperature in pyrolysis reactions to prevent carbon formation, improving performance, selectivity, and product separation. SFC can also be used in heterogeneous catalytic reactions that undergo catalyst deactivation as a result of coke deposition, in which coke deposits can be removed from the catalyst surface by means of SCF by reactivating the catalyst by removing low-volatility compounds. A supercritical medium can also be used in reactions where product separation using conventional techniques is difficult to achieve or very expensive. In these cases, the operating conditions can be adjusted to fractionate the reaction products. Other reasons for using SFC in chemical reactions are related to favorable mass transfer (viscosity, high diffusivity), enhanced reaction rates and process control [28].

scCO_2 has environmental (nontoxic, nonflammable), phase equilibrium (high solubility of volatile solvents, complete miscibility with gases), and chemical advantages that may be of great interest for its use as reaction medium. In this way, it can be noted that CO_2 cannot be oxidized, being interesting as a medium in oxidation reactions; generally, is not affected by free radicals, so it is an ideal solvent for reactions initiated by free radicals and is aprotic, so it can be used in cases where labile protons could interfere with the reaction.

Liquid and scCO_2 have been used in homogeneous and heterogeneous reactions, catalytic and noncatalytic, and in the synthesis and processing of polymers, including polymerization in homogeneous solution, precipitation polymerization, dispersion and polymerization in emulsion, and polycondensation [29].

4.3.1 Homogeneous catalytic reactions in supercritical carbon dioxide

Homogeneous catalysis may have certain advantages over heterogeneous catalysis, such as the possibility of carrying out the reaction under milder conditions, greater activity and selectivity, ease of spectroscopic monitoring, and controlled and tunable reaction sites. Although, heterogeneous catalytic reactions are of great importance in industries because of the easy separation of the catalyst after the reaction. Most homogeneous reactions carried out in scCO_2 are catalytic [30–33], while noncatalytic, supercritical, homogeneous reactions are generally carried out in supercritical water [5].

For homogeneous reactions, the main problem is to find an effective, CO_2-soluble catalyst, which can be achieved by adding functional groups. scCO_2 is more effective when used with nonpolar, nonionic, and low-molecular-mass compounds. However, it can be used with insoluble compounds in scCO2 by adding cosolvents.

Homogeneous hydrogenation reactions in scCO_2 have also been developed. Jessop et al. [34] studied the hydrogenation of carbon dioxide to formic acid, methanol, and other organic substances. They described the route to formic acid based on the use of Rh organometallic catalyst in dimethyl sulfoxide and aqueous solvents and also in a supercritical mixture of carbon dioxide and hydrogen containing a ruthenium catalytic complex. They concluded that SCF represents a promising medium for homogeneous catalysis. Rhodium complexes have also been synthesized for homogeneous catalysis of 1octene in scCO_2 from the phosphine reaction containing polymeric ligands of fluoroacrylate with an Rh complex [35].

Liu and Xiao [31] presented in 2007 a review summarizing the achievements in homogeneous and heterogeneous catalytic reactions with transition metal complexes in nonconventional media, such as ionic liquids, $scCO_2$, and fluorinated carbons. The study includes hydrogenation, hydroformylation, carbonylation, Heck reactions, Suzuki and Stille couplings, Sonogashira reactions, allylic substitution, olefin metathesis, olefin epoxidation, and alcohol oxidation. The homogeneous hydrogenation of 1-butene to n-butane with a membrane reactor is presented for homogeneous catalysis in $scCO_2$ with in situ catalyst separation [36].

Other homogeneous catalytic reactions using $scCO_2$ are summarized in **Table 3**.

4.3.2 Heterogeneous catalytic reactions in supercritical carbon dioxide

Reactions have also been carried out in $scCO_2$ with heterogeneous catalysts. SCFs, used as solvents or reactants, provide advantages to control and enhance heterogeneous catalytic reactions. Among these advantages can be mentioned: (i) reactants and products can be in a single phase, avoiding solubility problems and mass transfer limitations of conventional gas-liquid and liquid/liquid interfaces; ii) improve the transfer of internal mass through the pores of the catalyst, often resulting in an increase in reaction rate and selectivity; iii) enhance diffusion rate in reactions controlled by external (fluid/particle) diffusion; (iv) improve heat transfer; (v) easier product separation; (vi) inhibit deactivation of the catalyst by dissolution deposits, (vii) tune the solvent properties by changing pressure, temperature, and adding cosolvents; and (viii) thermodynamic pressure effect on rate constants [37].

$scCO_2$ is the most widely used as a solvent in heterogeneous catalysis, probably because reactions in $scCO_2$ have a behavior very similar to the reactions in nonpolar organic solvents (for example, n-hexane), and in practice, the replacement of these

Reaction	Catalyst
Hydrogenation	Ru, Rh complexes
Oxidation	Fe, Mn, Rh, Pt-Ru, Mo complexes, $PdCl_2$, $CuCl_2$
Dehydrogenation	Rh complex
Coupling	Pd, Ru complexes
Diels-Alder	Sc, Yb, Li complexes, $AlCl_2$
Esterification	Yb, Pd complexes, $PdCl_2$, $CuCl_2$
Alkylation	$CF_3(CF_2)_7SO_3H$
Hydroformulation	Rh, Ru, Co complexes
Carbonation	Re, Li, Sn, Zn, Fe complexes
Carbonylation	Pd, Ru complexes, $PdCl_2$, $CuCl_2$
Cyclization	Ni, Co complexes, $PdCl_2$, $CuCl_2$
Polymerization	Pd, Rh, Co, Sn, Ti complexes
Copolymerization	Zn, Cr, Al complexes
Epoxidation	Mo, V, Ti complexes
Transesterification	BF_3

Table 3.
Homogeneous catalytic reactions in supercritical solvents. Adapted from reference [5] with permission of Elsevier.

solvents by $scCO_2$ does not imply important changes, from the chemical point of view. However, this fact is also a limitation, since it is not suitable for hydrophilic substances and ions. Some authors, such as Johnston and Haynes [38], have raised the possibility of overcoming these limitations, for example, by using a water microemulsion, which is kept stable by adding a surfactant, in a continuous phase of carbon dioxide.

Many heterogeneous catalytic hydrogenation reactions that have been developed using $scCO_2$ supported Pt, Pd, Ni, Mo as catalysts. Some examples are cited below. Tacke et al. [39] carried out total and partial hydrogenation of fats and oils, free fatty acids, and fatty acid esters, using supercritical CO_2 as a solvent. They used a continuous fixed-bed reactor with a palladium catalyst on a commercial support and obtained yields up to six times higher than in the conventional hydrogenation process with palladium catalyst supported by activated carbon. Bertucco et al. [40] studied the catalytic hydrogenation of an unsaturated ketone in $scCO_2$ using a recycle reactor with a supported palladium catalyst. More recently, catalysts of supported noble metals, such as Ru, Pd, or Pt, have exhibited high activity for the hydrogenation of bio-based carboxylic acids. Leviulinic acid and succinic acid are converted into lactones or diols depending on the nature of the catalyst and reaction conditions [41].

Other heterogeneous catalytic reactions using $scCO_2$ are summarized in **Table 4**.

4.3.3 Supercritical biocatalysis

The use of SCF as an alternative to conventional organic solvents in enzyme-catalyzed reactions was first investigated by Randolph et al. [42], Hammond et al. [43], and Nakamura et al. [44] in 1985. Since then, it has been a very fruitful area of research. The possibility of modifying the physical properties of the solvent by changing the pressure or temperature offers great versatility [45, 46]. The tunability of the solvent is perhaps the main characteristic of biocatalysis in SCF. In addition, the activity of enzymes in nonaqueous media depends on the properties of the solvent, so SCFs are attractive media to carry out biocatalytic reactions [47]. The low viscosities and gas-like diffusivities improve the rate of mass transfer of reactants to the active sites in enzymes disseminated in SCF, in which enzymes are insoluble. Therefore, reactions limited by diffusion rates instead intrinsic kinetic will be faster in SCF than in liquids. In addition, higher substrate concentrations can also increase the observed reaction rates and improve process yields. The density of a supercritical fluid is sensitive to temperature and pressure, so that small changes in these variables result in significant changes in density and, consequently, in the properties of the solvent

Reaction	**Catalyst**
Hydrogenation	Supported Pt, Pd, Ni, Mo
Oxidation	Supported Pt, Pd, Ni, Co, Cu, Mn, Fe
Isomerization and rearrangement	Supported Pt ($scCO_2$ + cosolvent)
Cracking	Zeolite
Alkylation	Zeolite
Esterification	Biocatalysts, Zeolite
Hydroformulation	Supported Rh, Fe catalysts

Table 4.
Heterogeneous catalytic reactions in supercritical solvents. Adapted from reference [5] with permission of Elsevier.

that are a function of density, such as the solubility parameter, the dielectric constant, and the partition coefficient [48, 49]. Such changes in properties have been frequently studied for many solvents, then the experimental environment of the biocatalytic reaction can be reasonably controlled. However, despite the advantages of SCFs over organic solvents to carry out enzymatic reactions, their use is not as widespread as might be expected.

Due to its high availability and low toxicity, as has been already mentioned, the most widely used supercritical fluid is carbon dioxide, although its use in many processes is limited by the low solubility of many reactants and products, even in supercritical conditions. Many industrial applications are hampered by this handicap and by the cost of high-pressure equipment. Despite these drawbacks, the combination of environment-friendly catalysts and solvents has attracted the attention of many researchers, and there is a huge volume of literature available on the activity, specificity, and stability of different enzymes in $scCO_2$ [44, 50–52]. Furthermore, enzymes are very specific catalysts that require mild reaction conditions to exert their activity, which reduces the synthesis of undesirable or secondary products, and its activity [53] and selectivity [54] can be tuned by modifying the pressure or temperature of the supercritical fluid.

4.3.3.1 Enzymatic reactions in supercritical carbon dioxide ($scCO_2$)

Most of the work published so far in the field of biocatalysis in SCF uses $scCO_2$ as a reaction medium for the advantages over other SCFs, such as low cost, environmentally benign nature, low toxicity, high availability, nonflammability, and low critical temperature. Furthermore, using $scCO_2$ for enzymatic reactions, the extraction of the product is not necessary since the $scCO_2$ becomes gas with depressurization to atmospheric pressure. However, other SCFs, such as fluoroform and ethane, have been shown to be even more suitable as reaction medium for some enzymatic catalysis reactions [47].

One of the first works with $scCO_2$ was carried out by Randolph et al. [42]. They found that the enzyme alkaline phosphatase was active and stable in a batch reactor using $scCO_2$ as solvent. The enzyme catalyzed the reaction of *p*-nitrophenyl disodium phosphate to obtain *p*-nitrophenol, but its activity was limited by the solubility of *p*-nitrophenyl disodium phosphate in $scCO_2$.

The enzymes mostly used with $scCO_2$ are lipases. Many authors have investigated a wide variety of reactions using this class of enzymes. Nakamura et al. [44, 55, 56] studied the acidolysis of triolein with stearic acid in $scCO_2$ in batch and continuous processes, using four lipases, one of them in free form and the rest immobilized on a support. All of them were stable in $scCO_2$, and the authors found the optimal conditions to achieve the better productivity.

$scCO_2$ has also been used to obtain optical isomers by chiral synthesis from a racemic mixture. Ikushima et al. [57] studied the transesterification of (±)-citronellol with oleic acid using the lipase *Candida cylindracea*. The researchers found that increasing pressure of $scCO_2$, the reaction rate increased, especially in the vicinity of the critical point. Around the critical point, the formation of the S ester was stereoselective. Endo et al. [58] also obtained chiral esters from secondary alcohols and short-chain fatty acids using two different immobilized lipases in $scCO_2$.

Almeida et al. [59] carried out the transesterification reaction of butyl acetate by *n*-hexanol in supercritical ethane, $scCO_2$, and high-pressure propane using immobilized *Candida antarctica* lipase B (Novozym 435). The activity of Novozyme 435 activity was found to be similar in supercritical ethane and compressed propane, but about 1 order of magnitude lower in carbon dioxide under the same operating conditions (35°C and 10,000 kPa). However, the reaction rate increased with temperature in $scCO_2$.

One of the most determining factors to successfully carry out a reaction in biocatalysis is the stability of the enzyme, and the conditions involved in the use of $scCO_2$ can aggravate the problem. It has been shown that, in most cases, pressures below 20 MPa do not denature enzyme [60]. However, many cases have also been described in $scCO_2$ systems in which enzyme inactivation occurs [61]. Therefore, the progress of biocatalytic reactions in $scCO_2$ requires the development of the stabilization techniques.

Immobilized *Candida Antarctica* lipase B (CALB) was successfully used as catalyst to synthesize butyl butyrate from butyl vinyl ester and 1-butanol in $scCO_2$ with excellent results. The catalytic behavior of the enzyme immobilized on an acrylic support was studied in a stirred tank reactor, showing that a decrease in both the water content and the $scCO_2$ density enhanced the synthetic activity and selectivity [3].

4.3.3.2 $scCO_2/H_2O$ biphasic systems

Matsuda et al. [62] studied a reaction using a partially purified *Geotrichum candidum* alcohol dehydrogenase in a biphasic $scCO_2/H_2O$ system and observed an inactivation of the enzyme due to the low pH in the aqueous layer because of the high density of CO_2 (H_2CO_3).

A pressurized biphasic CO_2/H_2O system with pyruvate decarboxylase has been used to catalyze a carboxylation reaction, and it has been observed that the enzyme loses 80% of its activity at 6 MPa [63]. In order to stabilize the enzyme, additives such as glycerol and trehalose have been used, and it has been immobilized on an ion exchange polymer. After the stabilization, the activity was maintained up to CO_2 pressurized at 11 MPa.

4.3.3.3 $scCO_2$/ionic liquid biphasic systems

The growing interest that these systems have aroused lies in the combination of the excellent properties of ionic liquids as solvents to carry out numerous reactions, together with the magnificent qualities of $scCO_2$ as an extraction agent; so that it allows any reaction to be carried out in an ionic liquid and recover the products dissolved in CO_2 in a similar way, fast and clean (**Figure 19**).

Ionic liquids have also been used frequently with $scCO_2$ to improve the stability of the enzymes. It has been demonstrated that some enzymes have more native, stable, and compact conformation in $scCO_2$/ionic liquid system than in $scCO_2$ [62]. Reetz et al. [64] demonstrated the possibility of carrying out biocatalytic processes in biphasic systems based on ionic liquids and SCF. As in other catalytic reactions, the success of this type of biphasic systems is based on the high solubility of $scCO_2$ in the ionic liquid phase, while it does not present detectable solubility in the $scCO_2$ phase and in the demonstrated fact that $scCO_2$ can extract organic substances from the ionic liquid, without contamination of the extract by the latter [65], while the enzyme, free or immobilized, can be separated from the ionic liquid by a filtration process.

The development of technologies that directly provide pure products by the integration of reaction/separation processes, product recovery, and reuse of catalyst and solvents constitutes one of the objectives of green chemical engineering. The unique

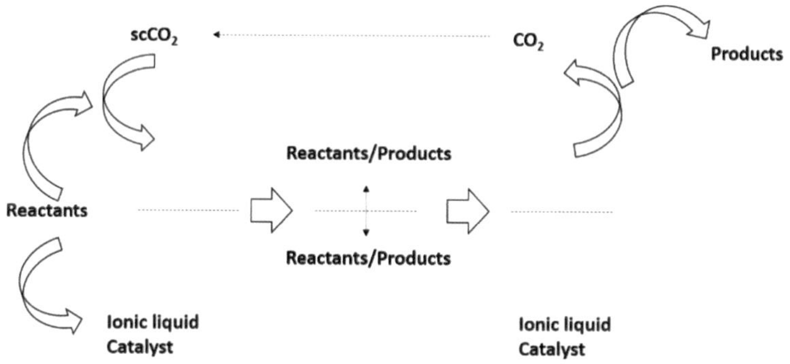

Figure 19.
Scheme of a reaction in a biphasic ionic liquid/scCO$_2$ system.

properties of ionic liquids lead, when combined with (bio)catalysts and $scCO_2$, to improvements, not only in catalytic efficiency (activity, enantioselectivity, stability, etc.), but also in the design of integrated processes for product separation (e.g., ionic liquid/$scCO_2$ biphasic reactors, membrane reactors, nanodrop systems, microfluidic devices, supported ionic liquid phases, sponge-like ionic liquids, etc.) [66]. These systems can also operate in semicontinuous systems, consisting of carrying out the reaction within the ionic liquid, while constantly passing CO_2 through the system to extract the products and collect them in a cold trap [64]. Biocatalytic transformations have been carried out under flow conditions in ionic liquids and SCFs as alternative nonaqueous reaction media. Several examples are provided (e.g., KR and DKR of *sec*-alcohols and amines, C–C bond formation, reduction, transamination, trans-esterification, etc.) where the use of continuous-flow techniques enables the development of more efficient processes and multiple reaction steps to be combined into a single continuous operation [67].

Other works [68] have combined ionic liquids, $scCO_2$, and membrane technology to perform a biocatalytic process, i.e., the synthesis of butyl propionate, from vinyl propionate and 1-butanol, in a membrane bioreactor with recirculation in a biphasic system ionic liquid/$scCO_2$, using ceramic tubular membranes of α-microporous alumina in which *Candida Antarctica* lipase was immobilized (**Figure 20**).

Recently, a chemo-enzymatic synthesis of omega-3 monoacylglycerides was carried out in two consecutive catalytic steps; first, an enzymatic transesterification of raw fish or linseed oil with solketal for producing fatty acid solketal esters, followed by the hydrolysis of these solketal moieties catalyzed by solid acids (e.g., zeolites) in either $scCO_2$ or sponge-like ionic liquids [69].

4.4 Other applications

SCFs can be also useful for applications related to the microelectronics processing, the dyeing of textiles, dry cleaning or washing of textile garments or machine parts; although, for now, this implies certain drawbacks by having to work discontinuously. Regarding this field of application, the main efforts are being focused on getting the parts in and out of a pressure vessel without having to open it. [70].

The treatment of toxic waste (where supercritical water can be used instead of incineration) and the remediation of soils are also considered as possible applications, although less widespread [71].

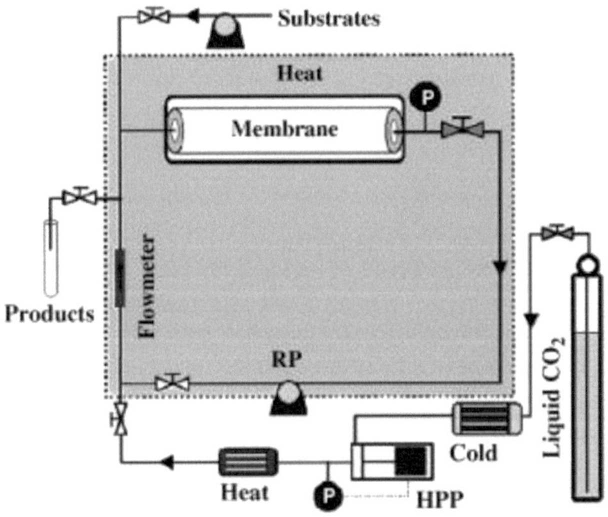

Figure 20.
Experimental setup of the recirculating enzymatic reactor with ionic liquid/scCO$_2$ system. From reference [68] with permission of Elsevier.

scCO$_2$ has been used as an environmentally friendly medium to deliver coating systems. Significant reductions in volatile organic compounds emission can be achieved by partially or totally replacing the organic solvent with CO$_2$ in spray coatings. In addition to spray coatings on different substrates, CO$_2$ processes can be used for powder coatings and also for the coating of preformed particles, such as metal powders and pharmaceuticals for controlled drug release [72] or for delivering biocides into wood [73].

Drug-loading implants have shown numerous advantages over simple implants. Impregnation in a scCO2 medium has been used to load drugs into polymeric implants, because it allows the recovery of the final implant, free of any solvent residue, and operate at a mild temperature that is suitable for processing with thermosensitive compounds [74].

Moreover, the sterilization of implantable medical devices is also of paramount importance to avoid complications related to surgery, such as infection and rejection. The use of scCO$_2$ for sterilization has recently been investigated. Aspects such as microorganisms that can be inactivated by scCO$_2$, operating variables, and materials sterilized by scCO$_2$ have been analyzed [75].

Recently, mold foaming process has been proposed to obtain thermoplastic polyurethane foam sheets using scCO$_2$, obtaining foams with good dimensional stability and cell morphology and excellent flexibility [76]. For applications of monolithic silica aerogels, the supercritical drying of silica aerogels using scCO$_2$ to substitute the liquid by gas has opened the possibility for a large-scale aerogel drying process [77].

Finally, another important field of research developed in the last two decades can be highlighted is the integration of SCF and ionic liquids. Reports have effectively shown the potential for combining ionic liquid and scCO$_2$ systems for product extraction and separation as well as reaction media [78].

Regarding the type of operation suitable for all these applications, it can be stated that in processes based on SCF, continuous operation is clearly advantageous compared with operation in stages. This is due to small equipment, which is much cheaper, can be used while maintaining high productivity. In the particular case of caffeine, the industrial process works semicontinuously with two columns, i.e., one is extracting while the other is being emptied and refilled again with coffee beans.

5. Conclusions

SCFs are fluids at a pressure and temperature conditions above their critical point. Under these conditions, these substances have very interesting properties as solvents (gas-like diffusivities and densities close to liquids), which can be modulated both by the environmental conditions (P and T) and by the presence of additional cosolvents for use as solvents and as reaction media. The most used supercritical fluid is carbon dioxide, since it has relatively low critical parameters (Pc = 72.8 bar, Tc = 31.5°C), with characteristics as a solvent close to organic compounds (modulable hydrophobicity), although it has different advantages over organics as solvent and reaction medium, such as high rates of matter transfer, high diffusion coefficients, zero toxicity, non flammability, and low cost, among others. In addition, supercritical carbon dioxide ($scCO_2$) is especially advantageous, as it is cheap, nontoxic, and nonflammable and has a critical temperature close to room temperature and moderate critical pressure and can be reused. When used as a reaction medium, the products obtained can be easily separated since the solubility of the same in the $scCO_2$ medium is a function of the pressure and temperature conditions of the $scCO_2$, so its solvent capacity can be modulated. All these properties make scCO2 very attractive to be used as a "green design solvent." As has been shown in this chapter, applications in different types of operation (extraction, reaction, obtaining nanoparticles, drying ...) and in different sectors (food and beverages, pharmaceutical, biomedical, microelectronics, textiles, forest products, petrochemicals, chemicals, environmental cleaning, production of synthetic fuels, polymers, coatings,...) have been growing since the 1980s.

Author details

Mercedes G. Montalbán* and Gloria Víllora
Chemical Engineering Department, University of Murcia, Faculty of Chemistry, Regional Campus of International Excellence "Campus Mare Nostrum", Murcia, Spain

*Address all correspondence to: mercedes.garcia@um.es

References

[1] Anderson MH, Ensher JR, Matthews MR, Wieman CE, Cornell EA. Observation of Bose-Einstein condensation in a dilute atomic vapor. Science. 1995;**269**(5221):198-201

[2] Sapkale GN, Patil SM, Surwase US, Bhatbhage PK. Supercritical Fluid Extraction: A Review. International Journal of Chemical Sciences. 2010;**8**(2):729-743

[3] Lozano P, Víllora G, Gómez D, Gayo AB, Sánchez-Conesa JA, Rubio M, et al. Membrane reactor with immobilized Candida antarctica lipase B for ester synthesis in supercritical carbon dioxide. Journal of Supercritical Fluids. 2004;**29**(1-2):121-128

[4] Aionicesei E, Škerget M, Knez Ž. Supercritical CO2, a clean alternative to traditional methods of processing polymeric biomaterials. Acta Medico-Biotechnica. 2009;**2**(2):19-24

[5] Sunol A, Sunol S, Cogswell K. SUPERCRITICAL SOLVENTS. In substitution of solvents by safer products. In: Wypych G, editor. Handbook of Solvents. 3rd Edition. Canada: ChemTec Publishing; 2019. pp. 1455-1634

[6] Weissermel K, Arpe H-J. Industrial Organic Chemistry, 3rd, Completely Revised Edition. Germany: Wiley; 2008. p. 481

[7] Seki T, Grunwaldt J-D, Baiker A. Heterogeneous catalytic hydrogenation in supercritical fluids: Potential and limitations. Industrial and Engineering Chemistry Research. 2008;**47**(14):4561-4585

[8] Liu L-J, Guo H-M, Xue B, Lou H, Chen M. Hydrogenation in supercritical conditions catalyzed by palladium supported on modified activated carbon. RSC Advances. 2015;**5**(82):66704-66710

[9] Castillo E, Marty A, Combes D, Condoret J. Polar substrates for enzymatic-reactions in supercritical CO2 - how to overcome the solubility limitation. Biotechnology Letters. 1994;**16**(2):169-174

[10] Coelho JP, Filipe RM, Paula Robalo M, Boyadzhieva S, St CG, Stateva RP. Supercritical CO2 extraction of spent coffee grounds. Influence of co-solvents and characterization of the extracts. Journal of Supercritical Fluids. 2020;**161**:104825

[11] Fischer M, Jefferies TM. Optimization of nicotine extraction from tobacco using supercritical fluid technology with dynamic extraction modeling. Journal of Agricultural and Food Chemistry. 1996;**44**(5):1258-1264

[12] Yoshida VMH, Balcão VMCF, Vila MMDC, Júnior JMO, Aranha N, Gremião MPD, et al. Supercritical fluid and pharmaceutical applications. Part I: Process classification. African Journal of Pharmaceutical Pharmacology. 2016;**10**(9):132-144

[13] Moldão-Martins M, Palavra A, Beirão Da Costa ML, Bernardo-Gil MG. Supercritical CO2 extraction of thymus zygis L. subsp. sylvestris aroma. Journal of Supercritical Fluids. 2000;**18**(1):25-34

[14] Donohue M, Geiger JL, Kiamos AA, Nielsen KA. Reduction of volatile organic compound emissions during spray painting: A new process using supercritical carbon dioxide to replace traditional paint solvents. ACS Symposium Series. 1996;**626**:152-167

[15] Kehili M, Kammlott M, Choura S, Zammel A, Zetzl C, Smirnova I, et al. Supercritical CO2 extraction and antioxidant activity of lycopene and β-carotene-enriched oleoresin from tomato (Lycopersicum esculentum L.) peels by-product of a Tunisian industry. Food and Bioproducts Processing. 2017;**102**:340-349

[16] Kazarian SG. Polymer processing with supercritical fluids. Polymer Science Series CC/C of vysokomolekuliarnye soedineniia. 2000;**42**(1):78-101

[17] Brunner G. Mass separation with supercritical gases (gas extraction). International Chemical Engineering. 1990;**30**(2):191-205

[18] Zabot GL. Chapter 11 - decaffeination using supercritical carbon dioxide. In: Inamuddin, Asiri AM, Isloor AMBT-GSP for C and EE and S, editors. Green Sustainable Process for Chemical and Environmental Engineering and Science. 1st Edition. Netherlands: Elsevier; 2020. p. 255-278

[19] McHugh MA, Krukonis VJ. In: Brenner H, editor. Supercritical Fluid Extraction: Principles and Practice. 2nd ed. Stoneham, MA, USA: Butterworth-Heinemann; 1994. p. 608

[20] Brunner G. Gas Extraction: An Introduction to Fundamentals of Supercritical Fluids and the Application to Separation Processes. Germany: Springer; 1994. p. 387

[21] Martín Á, Cocero MJ. Supercritical Fluids. In: Vol. Processes and Process Engineering. Kirk-Othmer Encyclopedia of Chemical Technology. 2016:1-28

[22] Perrut M. Supercritical fluid applications: Industrial developments and economic issues. Industrial and Engineering Chemistry Research. 2000;**39**(12):4531-4535

[23] Yeo S-D, Kiran E. Formation of polymer particles with supercritical fluids: A review. Journal of Supercritical Fluids. 2005;**34**(3):287-308

[24] York P, Hanna M. Particle engineering by supercritical fluid technologies for powder inhalation drug delivery in respiratory drug delivery. In: V Proceedings of the Conference on Respiratory Drug Delivery. Phoenix, Arizona; 1996. pp. 231-239

[25] Kankala RK, Chen B-Q, Liu C-G, Tang H-X, Wang S-B, Chen A-Z. Solution-enhanced dispersion by supercritical fluids: An ecofriendly nanonization approach for processing biomaterials and pharmaceutical compounds. International Journal of Nanomedicine. 2018;**13**:4227-4245

[26] Martín Á, Weidner E. PGSS-drying: Mechanisms and modeling. Journal of Supercritical Fluids. 2010;**55**(1):271-281

[27] Reverchon E. Supercritical-assisted atomization to produce micro- and/or nanoparticles of controlled size and distribution. Industrial and Engineering Chemistry Research. 2002;**41**(10):2405-2411

[28] Subramaniam B, McHugh MA. Reactions in supercritical fluids - a review. Industrial and Engineering Chemistry Process Design and Development. 1986;**25**(1):1-12

[29] Cooper AI. Polymer synthesis and processing using supercritical carbon dioxide. Journal of Materials Chemistry. 2000;**10**(2):207-234

[30] Savage PE, Gopalan S, Mizan TI, Martino CJ, Brock EE. Reactions at supercritical conditions: Applications and fundamentals. AICHE Journal. 1995;**41**(7):1723-1778

[31] Liu S, Xiao J. Toward green catalytic synthesis-transition metal-catalyzed reactions in non-conventional media. Journal of Molecular Catalysis A: Chemical. 2007;**270**:1-43

[32] Corma A, García H. Lewis acids: From conventional homogeneous to green homogeneous and heterogeneous catalysis. Chemical Reviews. 2003;**103**(11):4307-4366

[33] Skouta R. Selective chemical reactions in supercritical carbon dioxide, water, and ionic liquids. Green Chemistry Letters and Reviews. 2009;**2**(3):121-156

[34] Jessop PG, Ikariya T, Noyori R. Homogeneous catalytic hydrogenation of supercritical carbon dioxide. Nature. 1994;**368**(6468):231-233

[35] Kani I, Omary MA, Rawashdeh-Omary MA, Lopez-Castillo ZK, Flores R, Akgerman A, et al. Homogeneous catalysis in supercritical carbon dioxide with rhodium catalysts tethering fluoroacrylate polymer ligands. Tetrahedron. 2002;**58**(20):3923-3928

[36] Goetheer ELV, Verkerk AW, Van Den Broeke LJP, De Wolf E, Deelman BJ, Van Koten G, et al. Membrane reactor for homogeneous catalysis in supercritical carbon dioxide. Journal of Catalysis. 2003;**219**(1):126-133

[37] Baiker A. Supercritical fluids in heterogeneous catalysis. Chemical Reviews. 1999;**99**(2):453-474

[38] Johnston KP, Haynes C. Extreme solvent effects on reaction rate constants at supercritical fluid conditions. AICHE Journal. 1987;**33**(12):2017-2026

[39] Tacke T, Wieland S, Panster P. Hardening of fats and oils in supercritical CO2. In: von Rohr PR, Trepp CBT-PTP, editors. High Pressure Chemical Engineering. Netherlands: Elsevier; 1996. p. 17-21

[40] Bertucco A, Canu P, Devetta L, Zwahlen AG. Catalytic hydrogenation in supercritical CO2: Kinetic measurements in a Gradientless internal-recycle reactor. Industrial and Engineering Chemistry Research. 1997;**36**(7):2626-2633

[41] Corbel-Demailly L, Ly B-K, Minh D-P, Tapin B, Especel C, Epron F, et al. Heterogeneous catalytic hydrogenation of biobased levulinic and succinic acids in aqueous solutions. ChemSusChem. 2013;**6**(12):2388-2395

[42] Randolph TW, Blanch HW, Prausnitz JM, Wilke CR. Enzymatic catalysis in a supercritical fluid. Biotechnology Letters. 1985;**7**(5):325-328

[43] Hammond DA, Karel M, Klibanov AM, Krukonis VJ. Enzymatic reactions in supercritical gases. Applied Biochemistry and Biotechnology. 1985;**11**(5):393-400

[44] Nakamura K, Chi Y, Yamada Y, Yano T. Lipase activity and stability in supercritical carbon dioxide. Chemical Engineering Communications. 1986;**45**(1-6):207-212

[45] Kamat SV, Beckman EJ, Russell AJ. Enzyme activity in supercritical fluids. Critical Reviews in Biotechnology. 1995;**15**(1):41-71

[46] Chulalaksananukul W, Condoret J-S, Combes D. Geranyl acetate synthesis by lipase-catalyzed transesterification in supercritical carbon dioxide. Enzyme and Microbial Technology. 1993;**15**(8):691-698

[47] Mesiano AJ, Beckman EJ, Russell AJ. Supercritical Biocatalysis. Chemical Reviews. 1999;**99**(2):623-634

[48] Laane C, Boeren S, Hilhorst R, Veeger C. Optimization of biocatalysis in organic media. In: Laane C, Tramper J, Lilly MD, editors. Biocatalysis in Organic Media. 1st ed. Amsterdam: Elsevier Science Publishers; 1987. pp. 65-84

[49] Kamat SV, Iwaskewycz B, Beckman EJ, Russell AJ. Biocatalytic synthesis of acrylates in supercritical fluids: Tuning enzyme activity by changing pressure. Proceedings of the National Academy of Sciences of the United States of America. 1993;**90**(7):2940-2944

[50] Randolph TW. Supercritical fluid extractions in biotechnology. Trends in Biotechnology. 1990;**8**(3):78-82

[51] Paulaitis ME, Krukonis VJ, Kurnik RT, Reid RC. Supercritical fluid extraction. Reviews in Chemical Engineering. 1983;**1**(2):179-250

[52] Marty A, Chulalaksananukul W, Condoret JS, Willemot RM, Durand G. Comparison of lipase-catalysed esterification in supercritical carbon dioxide and in n-hexane. Biotechnology Letters. 1990;**12**(1):11-16

[53] Randolph TW, Clark DS, Blanch HW, Prausnitz JM. Cholesterol aggregation and interaction with cholesterol oxidase in supercritical carbon dioxide. Proceedings of the National Academy of Sciences of the United States of America. 1988;**85**(9):2979-2983

[54] Kamat SV, Beckman EJ, Russell AJ. Control of enzyme enantioselectivity with pressure changes in supercritical fluoroform. Journal of the American Chemical Society. 1993;**115**(19):8845-8846

[55] Nakamura K. Biochemical reactions in supercritical fluids. Trends in Biotechnology. 1990;**8**:288-292

[56] Rizvi SSH. Supercritical Fluid Processing of Food and Biomaterials. England: Blackie Academic & Professional; 1994

[57] Ikushima Y, Saito N, Yokoyama T, Hatakeda K, Ito S, Arai M, et al. Solvent effects on an enzymatic Ester synthesis in supercritical carbon dioxide. Chemistry Letters. 1993;**22**(1):109-112

[58] Yano T, Matsuno R, Nakamura K. Developments in Food Engineering: Proceedings of the 6th International Congress on Engineering and Food. London; New York: Blackie Academic & Professional; 1994

[59] Almeida MC, Ruivo R, Maia C, Freire L, Corrêa de Sampaio T, Barreiros S. Novozym 435 activity in compressed gases. Water activity and temperature effects. Enzyme and Microbial Technology. 1998;**22**(6):494-499

[60] Mozhaev VV, Heremans K, Frank J, Masson P, Balny C. High pressure effects on protein structure and function. Proteins. 1996;**24**(1):81-91

[61] Harada T, Kubota Y, Kamitanaka T, Nakamura K, Matsuda T. A novel method for enzymatic asymmetric reduction of ketones in a supercritical carbon dioxide/water biphasic system. Tetrahedron Letters. 2009;**50**(34):4934-4936

[62] Matsuda T. Recent progress in biocatalysis using supercritical carbon dioxide. Journal of Bioscience and Bioengineering. 2013;**115**(3):233-241

[63] Matsuda T, Nakayama K, Abe T, Mukouyama M. Stabilization of pyruvate decarboxylase under a pressurized carbon dioxide/water biphasic system. Biocatalysis and Biotransformation. 2010;**28**(3):167-171

[64] Reetz MT, Wiesenhöfer W, Franciò G, Leitner W. Biocatalysis in

ionic liquids: Batchwise and continuous flow processes using supercritical carbon dioxide as the mobile phase. Chemical Communications. 2002;**9**:992-993

[65] Blanchard LA, Gu Z, Brennecke JF. High-pressure phase behavior of ionic liquid/CO2 systems. The Journal of Physical Chemistry. B. 2001;**105**(12):2437-2444

[66] Villa R, Alvarez E, Porcar R, Garcia-Verdugo E, Luis SV, Lozano P. Ionic liquids as an enabling tool to integrate reaction and separation processes. Green Chemistry. 2019;**21**(24):6527-6544

[67] Garcia-Verdugo E, Porcar R, Luis SV, Lozano P. Chapter 2 green biotransformations under flow conditions. In: Flow Chemistry: Integrated Approaches for Practical Applications. England: The Royal Society of Chemistry; 2020. pp. 50-85

[68] Hernández FJ, de los Ríos AP, Jiménez M, Rubio M, Gómez D, Víllora G. A new recirculating enzymatic membrane reactor for ester synthesis in ionic liquid/supercritical carbon dioxide biphasic systems. Applied Catalysis B: Environmental. 2006;**67**(1-2):121-126

[69] Villa R, Alvarez E, Nieto S, Donaire A, Garcia-Verdugo E, Luis SV, et al. Chemo-enzymatic production of omega-3 monoacylglycerides using sponge-like ionic liquids and supercritical carbon dioxide. Green Chemistry. 2020;**22**(17):5701-5710

[70] Beckman EJ. Supercritical and near-critical CO2 in green chemical synthesis and processing. Journal of Supercritical Fluids. 2004;**28**:121-191

[71] Veriansyah B, Kim J-D. Supercritical water oxidation for the destruction of toxic organic wastewaters: A review. Journal of Environmental Sciences (China). 2007;**19**(5):513-522

[72] Hay JN, Khan A. Review environmentally friendly coatings using carbon dioxide as the carrier medium. Journal of Materials Science. 2002;**37**(22):4743-4752

[73] Kjellow AW, Henriksen O. Supercritical wood impregnation. Journal of Supercritical Fluids. 2009;**50**(3):297-304

[74] Champeau M, Thomassin J-M, Tassaing T, Jérôme C. Drug loading of polymer implants by supercritical CO2 assisted impregnation: A review. Journal of Controlled Release. 2015;**209**:248-259

[75] Soares GC, Learmonth DA, Vallejo MC, Davila SP, González P, Sousa RA, et al. Supercritical CO_2 technology: The next standard sterilization technique? Materials Science & Engineering. C, Materials for Biological Applications. 2019;**99**:520-540

[76] Jiang X, Zhao L, Feng L, Chen C. Microcellular thermoplastic polyurethanes and their flexible properties prepared by mold foaming process with supercritical CO2. Journal of Cellular Plastics. 2019;**55**(6):615-631

[77] García-González CA, Camino-Rey MC, Alnaief M, Zetzl C, Smirnova I. Supercritical drying of aerogels using CO2: Effect of extraction time on the end material textural properties. Journal of Supercritical Fluids. 2012;**66**:297-306

[78] Blanchard LA, Hancu D, Beckman EJ, Brennecke JF. Green processing using ionic liquids and CO2. Nature. 1999;**399**(6731):28-29

Chapter 2

High-Pressure Fluid Phase Equilibria

Mercedes G. Montalbán and Gloria Víllora

Abstract

One of the crucial aspects in the design of processes of this millennium is the use of environmentally benign technologies. The introduction of supercritical fluids (SCF) and, in addition, their use with other solvents, such as ionic liquids, further diversify the applications of these fluids. SCF are powerful solvents with many unique properties. They have the mobility of gases and the dissolving power of liquid solvents, resulting in efficient high mass transfer rates and penetration into porous matrices. However, reliable and versatile mathematical models of phase equilibrium thermodynamics are needed for use in process design and viability studies. This chapter reviews experimental procedures for obtaining high-pressure phase equilibria data. In addition, phase diagrams describing binary mixtures and thermodynamic models capable of determining the conditions at phase equilibria at high pressures are considered.

Keywords: critical point, phase equilibria, experimental, synthetic, analytical, thermodynamic modeling, equations of state, mixing rules

1. Introduction

Phase behavior of mixtures is essential in supercritical fluid (SCF) process design. In these processes, feasibility and optimal conditions can only be established if phase equilibrium and solubility data are available. For this reason, there are many studies that have been carried out to elucidate the phase behavior in systems involving SCF and, particularly, carbon dioxide + solute systems. Numerous experimental methods for investigating high-pressure phase equilibria have been described and reviews since the 80s to the present day have been complicated. These reviews include the experimental method used and data from vapor-liquid equilibria (VLE), liquid-liquid equilibria (LLE), vapor-liquid-liquid equilibria (VLLE), solubility of high boiling point substances in SCF, and gas solubility in liquids [1–6].

Although the effectiveness of SCF such as supercritical water and supercritical carbon dioxide in variety of process is very promising, any industrial supercritical application to be designed depends on the possibility of modeling and predicting the phase equilibria in the systems involved. Only through this knowledge, engineers will be able to estimate feasibility and choose conditions to optimize the process. In the same way, processes with little chance of success can be identified and rejected. Therefore, the design of supercritical processes requires a sufficiently detailed understanding of the actual molecular process in SCF mixtures, and using this knowledge to search for correlations and develop reliable, versatile predictive models. However, this task is not easy. There are some aspects of SCF behavior that make them especially difficult to handle in the lab and to model. One of them is derived from the operating conditions to achieve the supercritical state, the other is the proximity of the conditions to the critical point, and another, the asymmetry of most of the SCF systems of interest is in terms of size and the attractive forces of the molecules involved.

The purpose of this chapter is to achieve a brief review of the experimental and analytical procedures in phase equilibrium thermodynamics of SCF systems.

2. Experimental acquisition of high-pressure fluid phase equilibrium data

This section consists of a brief review of the techniques for measuring the high-pressure phase equilibria. A large number of experimental methods have been developed for the measurement of the phase equilibrium of fluids at high pressure [5]. This is due to high-pressure phase behavior is often complex and difficult to predict, and hence, no single method is appropriate to study all different systems. As a general classification, the used methods can be divided into analytical or direct methods, and synthetic or indirect methods, depending on how the composition of the phases is measured. Some extensive reviews on the experimental methods used in previous works can be found in the literature [3–5].

2.1 Analytical methods

Analytical methods involve the analytical determination of the compositions of the co-existing phases. They can be subdivided into static, recirculating, and flow methods, depending on the technique used to achieve phase equilibrium.

The *static method* is represented in **Figure 1** [7]. A liquid-vapor mixture is inside the cell, whose pressure and temperature are controlled. A stirring system is normally used to facilitate phase equilibrium. Once this equilibrium is reached, small samples are taken from both the vapor phase and the liquid phase and their composition is analyzed. The main drawback of this method is that the sampling can alter the equilibrium of the phases. Static analytical methods have been widely used in the literature for measuring the phase equilibria [8–10].

The second analytical method is the *recirculation method*, which is shown in **Figure 2**. In this method, the liquid phase and the vapor phase are recirculated in parallel flow to achieve a better mixing between phases and to guarantee that the phase equilibrium is reached. It uses a closed cell similar to that of the static method. The composition of the phases is determined by collecting the corresponding samples through the sampling valves that are on-line connected to an analysis equipment. Then, main drawbacks of this method are that undesirable pressure gradients across the equilibrium cell can be provoked by the circulation pump and the need for a uniform temperature field to avoid partial condensation or vaporization in the recirculation lines [5]. This method has been widely described in previous works [11–13].

In the *flow method*, which is represented in **Figure 3**, high-pressure pumps are in charge of feed the preheated components into a static mixer where the equilibrium is attained. The feed stream from the mixer is separated into a vapor and liquid phases in an equilibrium cell. A liquid or heavy phase is continuously withdrawn from the bottom of the equilibrium cell, while the vapor or light phase is withdrawn from the

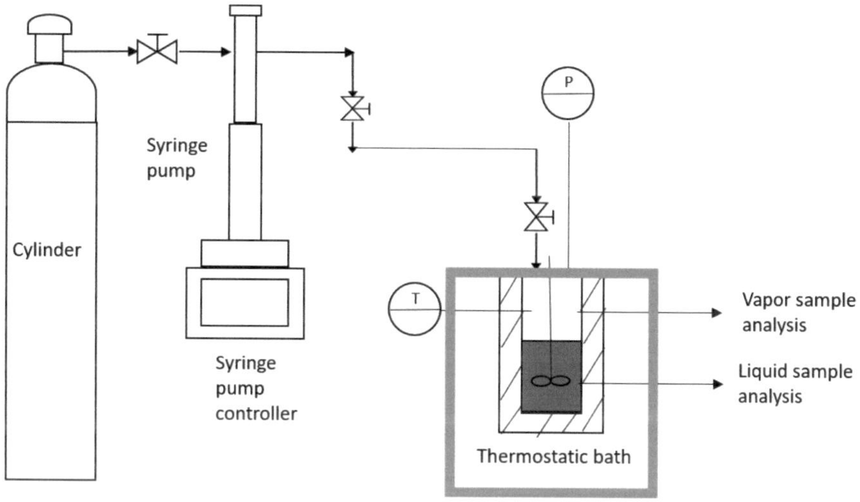

Figure 1.
Static method.

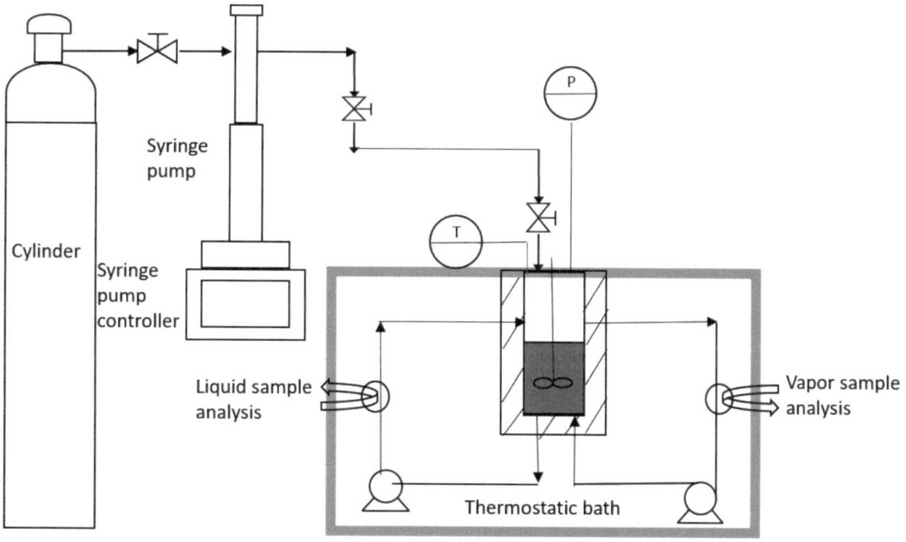

Figure 2.
Recirculation method.

top. Both phases are then depressurized, properly collected, and analyzed. Flow methods have the advantage that sampling does not alter the equilibrium. In addition, large quantities of sample for analysis can be generated when components are in very low concentrations because the run time of the experiment can be extended to accumulate more material. This method is preferred when working with compounds that are temperature-sensitive due to the short residence time of the components in the equipment. As a drawback, this method can be only used for systems that need short times to achieve phase equilibrium [5]. Several authors have used flow methods to obtain equilibrium data [14–16].

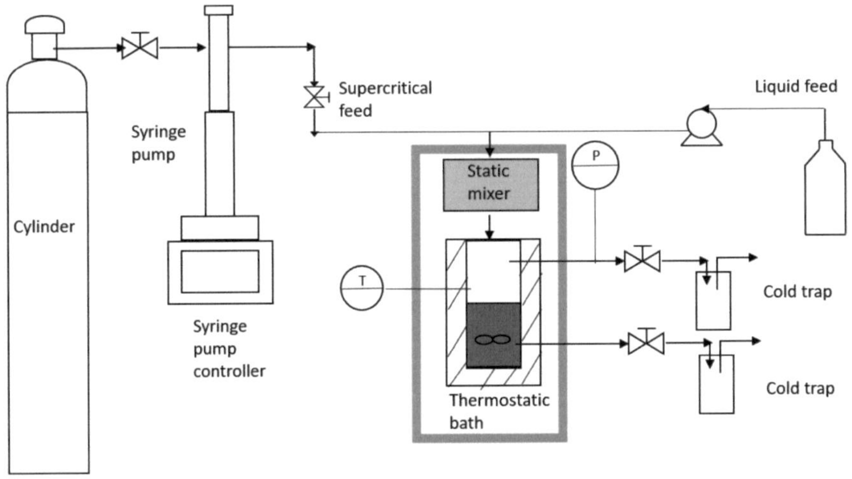

Figure 3.
Flow method.

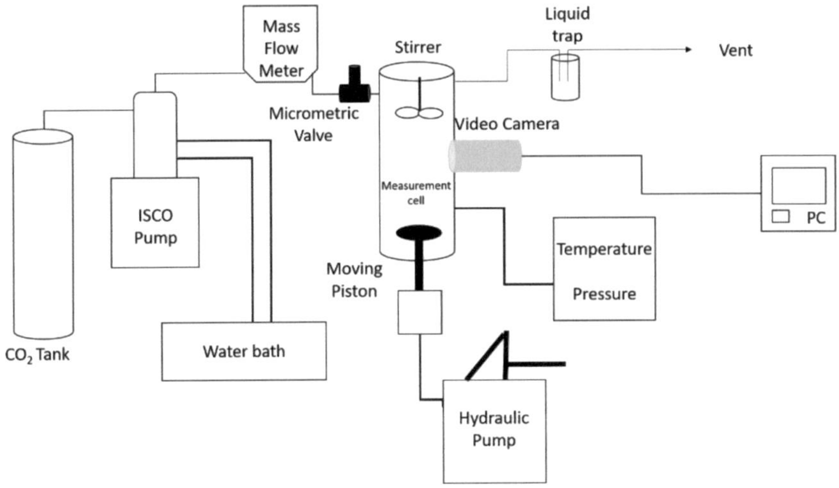

Figure 4.
Scheme of a possible experimental device for a synthetic method.

2.2 Synthetic methods

Synthetic methods are based on the preparation of a mixture of precisely known composition, the observation of the phase behavior in an equilibrium cell, and the measurement of the properties in the equilibrium state, for example, pressure and temperature. Therefore, no sampling is necessary and hence, no alteration of the equilibrium can occur [5]. **Figure 4** shows an example of the experimental device.

Synthetic methods can be used for systems with or without phase transition:

- *Synthetic methods with phase transition*: Temperature and pressure values are adjusted so that the mixture is homogeneous, that is, there is only a phase. After a variation of pressure or temperature, the formation of a new phase can be observed. The new phase can be detected by visual observation [17–19] or by changes in certain physical properties [20–22].

- *Synthetic methods without phase transition*: These methods are less frequent than synthetic methods with phase transition. Equilibrium properties such as pressure, temperature, phase volumes, and densities are measured and phase compositions are calculated using the material balance. They can be isothermal or isobaric [23–25].

Synthetic methods can be used when the analytical methods are not suitable. For instance, when the coexisting phases have a similar density, as occurs in the critical region [3]. However, visual observation of a new phase is difficult for those systems in which both phases have approximately the same refractive index [26]. This method is not practical for multicomponent systems because the tie lines cannot be determined without carrying out additional experiments [5]. The great advantages of synthetic methods are that the experimental procedures are usually easy and quick and that analytical equipment and complex sampling are not required.

3. Classification of the phase diagrams for binary mixtures

When a system consists of more than one substance, it is important to determine how the composition of the phases in equilibrium varies with temperature, pressure, and/or the initial composition of the system. These changes are key to carry out the well-known unit operations used industrially, such as distillation and extraction.

In isolated systems without applied external fields, composite systems tend to homogenize (although if the limit of stability is reached, separate phases of different composition are segregated, but always of homogeneous chemical potential in equilibrium). Whether a binary mixture remains a stable homogeneous fluid or is divided into two or more phases in equilibrium is determined by its thermodynamic stability. A mixture can be considered stable when its Gibbs- or Helmholtz-free energy is at a minimum.

In the case of two component systems, three variables are required to graphically represent the stability field of a homogeneous region (single-phase), which makes necessary a three-dimensional diagram. For convenience, a constant variable (pressure, temperature, or composition) is usually maintained and two-dimensional phase diagrams are represented, which are the cross sections of the three-dimensional representation.

The phase diagram to describe binary mixtures depends on the behavior of the species. At high pressures, a wide variety of phase behaviors can occur. Van Konynenburg and Scott [27] classified the behavior of the phases of binary mixtures into six types of pressure-temperature diagrams, considering van der Waals equation of state and quadratic mixing rules. **Figure 5** presents the different types of phase diagrams [28].

The Type I diagram (**Figure 5**) represents a phase behavior that frequently occurs when the two components of the mixture have critical properties of the same magnitude or a similar interaction energy and molecular size [29]. It is characterized by a continuous critical site between the two critical points of the two pure components and does not show any region of liquid-liquid immiscibility between the components [7]. An example of binary mixture showing type I behavior was found by Wei et al. [30] for the mixture of methane + ethane, in which both components are similar nonpolar molecules.

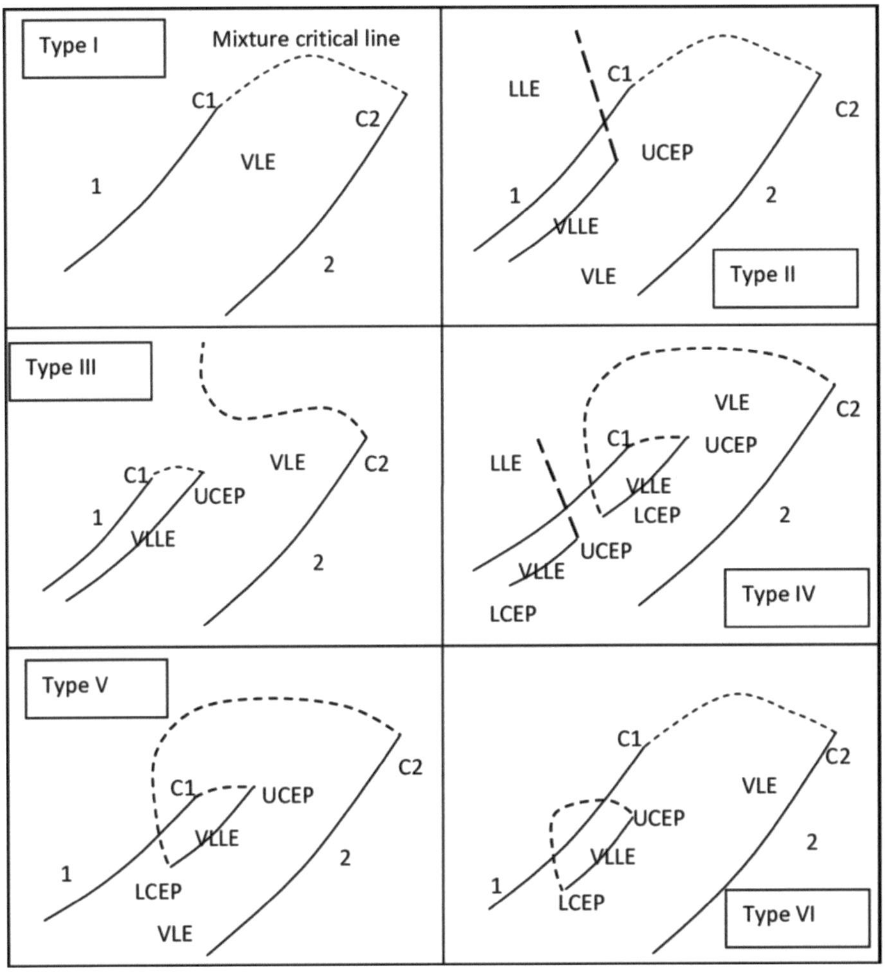

Figure 5.
Different types of phase behavior in binary fluid systems. C: Critical point, V: Vapor, L: Liquid, C1: Critical point of the most volatile component, C2: Critical point of least volatile component, LCEP: Lower critical end point. UCEP: Upper critical end point. Dashed curves are critical [28].

The Type II phase behavior is similar to Type I, but presents a zone of liquid-liquid immiscibility at low temperatures. As can be seen in **Figure 5**, there is a vapor-liquid-liquid (VLLE) line, along which the three phases (two liquids and one vapor) are in equilibrium, whose end is called the upper critical end point (UCEP). At that point, the two liquid phases come together and merge into a single liquid phase. The end of the liquid-liquid line is called the upper critical solution temperature (UCST), where the two liquids come together to form a single liquid phase when the system temperature increases. An example of this type of phase behavior can be found in the carbon dioxide-1-butanol system [31, 32]. The authors found that CO_2 solubility in 1-butanol decreases with increasing temperature, while increasing when pressure raises, although the solubility of 1-butanol in CO_2 is relatively low and shows no major changes as temperature or pressure varies, except in the vicinity of the critical point. It was found that vapor-liquid equilibrium exists in a large range of experimental conditions and that the liquid-liquid-vapor zone ends at an UCEP at 22.99 bar and 259.25 K.

Type III phase behavior is frequently found in mixtures of components with high immiscibility. In these systems, the liquid-liquid line rises to higher temperatures and eventually intersects with the vapor-liquid curve, resulting in a discontinuous critical curve. This critical curve begins at the critical point of the least volatile component C2 and spread to higher pressures, while the natural change from vapor-liquid to liquid-liquid occurs. The other critical curve runs from the critical point of the most volatile component C1 to the UCEP, in which the liquid and vapor phases are critically linked in a single fluid phase in the presence of another liquid phase. An example of the type III phase behavior is the carbon dioxide-1-hexanol mixture [33]. This binary system exhibits a phase behavior with three critical curves: a critical vapor-liquid curve that starts from the least volatile component, that is, 1-octanol, and continues with another liquid-liquid curve toward higher pressures, and the third, again a critical vapor-liquid curve, starting from the most volatile component, that is, CO_2, and ending at UCEP, where intersects with the three-phase equilibrium curve (VLLE).

Type IV and type V phase behaviors are quite similar to types II and I, respectively [34]. The main difference lies in the liquid-vapor critical line that no longer continuously joins the critical points of pure components. The critical line starting from pure component two ends in a three-phase line from which a second critical line arises, which is connected to the critical point of the lightest pure component 1. At the intersection between the three-phase line and the critical lines are two critical end points: The one with the lowest temperature is called the lower critical end point (LCEP) and another with the highest temperature UCEP. These phase behaviors usually occur when the critical properties of the two components are very different because there are large differences in structure, molecular size, or intermolecular forces. An example of Type IV phase behavior has been shown by Kodama et al. [35] for the mixture ethane +1-butanol. They studied the phase equilibria and saturated densities of this system at high pressures using a static circulation apparatus at 313.15 K. The CO_2 + nitrobenzene binary system can be mentioned as example of type V phase behavior. Hou et al. [36] studied this system in a high-pressure variable volume view cell using an analytical method. Phase boundaries were measured at temperatures of 298.15, 310.45, and 322.75 K under pressures between 2.76 and 12.83 MPa and they found that three-phase equilibria exist in a temperature range of 303.60 to 313.65 K. Experimental data could be correlated with the Peng-Robinson equation of state (PR EoS) and two binary parameters.

Type VI phase behavior is presented by some rare fluid mixtures. They are characterized because they show a three-phase line that begins in an LCEP and ends in a UCEP. The two critical end points are connected by a critical liquid-liquid curve showing an elliptical minimum-pressure critical point. This type of phase behavior is also characterized by a continuous vapor-liquid critical locus extending between the two critical points of pure components. The behavior of type VI is not derived from the Van der Waals equation so van Konynenburg and Scott [27] did not include it in their initial classification.

Konynenburg and Scott's classification was presented as a succession of unrelated phase diagram types; that is, a given binary system belongs to a single type (from I to VI) of fluid phase behavior. However, after the publication of Van der Waals' thesis, it is known that there is a continuous path between liquid and vapor; therefore, there must be a continuous path between the six types of phase behavior. Transitions from one type to another are often observed experimentally in homologous series. For example, systems (CO_2 + n-C12), (CO_2 + n-C13), and (CO_2 + n-C14) exhibit phase diagrams of type II, type IV and type III, respectively. Then, there is a relationship between the size difference and the interactions and the type of resulting phase behavior. These phase transitions can be reproduced by an equation of state and modeling homologous series; however, this is not the best option because the transition from one type to another would be observed for a number of carbon atoms, which would not be an integer. To obviate this problem, it is better to choose a well-defined binary system and vary the binary interaction parameter between the components. In this way, the interactions between the two molecules change and the transitions between types of phase behavior can be continuously observed [28].

Figure 6 illustrates the evolution of phase diagrams. The transition between each type of phase diagram can be explained by considering the size effects of molecules and the repulsive interactions between them.

4. Thermodynamic modeling of high-pressure fluid phase equilibria

If a liquid mixture is in equilibrium with a vapor mixture being both at the same temperature (T) and pressure (P), for every component i in a mixture the condition of thermodynamic equilibrium is given by the following expression [37]:

$$\hat{f}_i^V = \hat{f}_i^L \tag{1}$$

where $\hat{f}_i^V$ and $\hat{f}_i^L$ are the fugacities of the component i in the vapor and liquid phases, respectively.

Fugacity can be expressed in terms of the fugacity coefficient so Eq. (1) can be expressed also as follows:

$$y_i \hat{\phi}_i^V P = x_i \hat{\phi}_i^L P \tag{2}$$

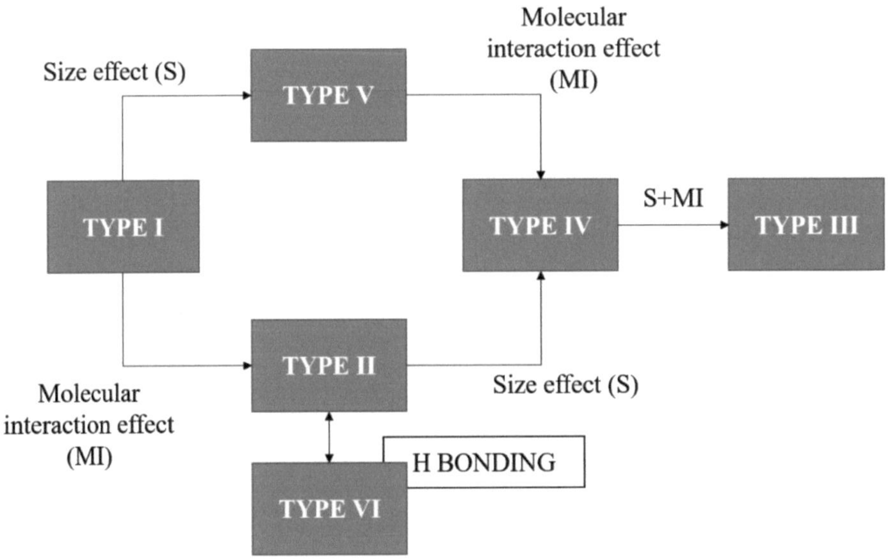

Figure 6.
Possible transitions between the different types of phase diagrams. Adapted from [28].

where y_i and x_i are the mole fractions of the component i in the vapor and liquid phase, respectively. The fugacity coefficients of the liquid and vapor phases can be calculated from the following thermodynamic expression [37]:

$$ln\,\hat{\phi}_i = \frac{1}{RT}\int_V^\infty \left[\left(\frac{\partial P}{\partial n_i}\right)_{T,V,n_{i\neq j}} - \frac{RT}{V}\right]dV - ln\left(\frac{Pv}{RT}\right) \tag{3}$$

where R is the gas constant, V is the total volume of the phase, v is the molar volume of the phase, and n_i and n_j are the moles of components i and j, respectively. The fugacities can be calculated from a simple equation of state.

4.1 Equations of state

An equation of state (EoS) is an algebraic relation between P, V, and T, which serves for describing some behavior of Nature. In this section, we will discuss the most common EoS models. The well-known van der Waals equation of state (vdW EoS) was proposed in 1873 and was the first equation able to represent liquid-vapor coexistence to binary mixtures [38]:

$$P = \frac{RT}{v-b} - \frac{a}{v^2} \tag{4}$$

The wdW EoS is a cubic EoS based on molecular interactions of attraction and repulsion. It can quantitatively predict most of the phase equilibrium behaviors exhibited by binary mixtures (39). However, the calculations are quantitatively inaccurate. Because of this, a large number of modifications to the vdW EoS have been developed, some of which (the most important) will be reviewed here.

For high-pressure phase equilibrium, the most widely used equations are the Peng-Robinson (PR) [40] and the Soave-Redlich-Kwong (SRK) [41] EoS.

The PR EoS has the following form for pure components:

$$P = \frac{RT}{v-b} - \frac{a[T]}{v(v+b)+b(v-b)} \tag{5}$$

where P, T, v, and R are the pressure, temperature, molar volume and the ideal gas constant, respectively, while $a[T]$ and b are the PR constants that are obtained from the critical temperature (T_c), critical pressure (P_c) and acentric factor (ω) as describe below:

$$a[T] = 0.45724\left(\frac{R^2T_c^2}{P_c}\right)\alpha[T_r] \tag{6}$$

$$T_r = \frac{T}{T_c} \tag{7}$$

$$\alpha[T_r] = \left(1 + m\left[1 - T_r^{0.5}\right]\right)^2 \tag{8}$$

$$m = 0.37464 + 1.54226\omega - 0.26992\omega^2 \tag{9}$$

$$b = 0.0778\left(\frac{RT_c}{P_c}\right) \tag{10}$$

The above equations closely reproduce vapor-liquid equilibria measurements for a number of different systems.

For its part, the SRK EoS is defined by the following equations:

$$P = \frac{RT}{v-b} - \frac{a[T]}{v(v+b)} \tag{11}$$

$$a[T] = 0.42747\left(\frac{R^2T_c^2}{P_c}\right)\alpha[T_r] \tag{12}$$

$$m = 0.48 + 1.574\omega - 0.176\omega^2 \tag{13}$$

$$b = 0.08664\left(\frac{RT_c}{P_c}\right) \tag{14}$$

where T_r and $\alpha[T_r]$ are calculated with Eqs. (7) and (8), respectively.

SRK and PR EoS are able to precisely model phase equilibrium based solely on the critical properties and acentric factors of the pure compounds. As a drawback, both EoS have limitations for predicting the densities of saturated liquids and low accuracy when working with highly polar systems.

Some years later, Stryjek and Vera (SV) [42] developed a variation of PR EoS by modifying the temperature-dependent function, $\alpha[T_r]$, to broaden the range of application to polar compounds. The variation is shown in Eq. (15):

$$m = \left(0.379 + 1.490\omega - 0.171\omega^2 + 0.020\omega^3\right) + \kappa_1\left(1 + \left(\frac{T}{T_c}\right)^{0.5}\right)\left(0.7 - \left(\frac{T}{T_c}\right)\right) \tag{15}$$

where κ_1 is an adjustable parameter, which is characteristic of each pure compound.

Patel and Teja (PT) also [43] proposed a new cubic EoS that reproduce the good features of the SRK and PR EoS for nonpolar fluids but, at the same time, overcomes some of the limitations of these equations for polar systems such as heavy hydrocarbons. The PT EoS is shown in Eq. (16):

$$P = \frac{RT}{v-b} - \frac{a[T]}{v(v+b)+c(v-b)} \tag{16}$$

where

$$a[T] = \Omega_a\left(\frac{R^2T_c^2}{P_c}\right)\alpha[T_r] \tag{17}$$

$$b = \Omega_b\frac{RT_c}{P_c} \tag{18}$$

$$c = \Omega_c\frac{RT_c}{P_c} \tag{19}$$

$$\Omega_c = 1 - \zeta_c \tag{20}$$

$$\Omega_a = 3\zeta_c^2 + 3(1-2\zeta_c)\Omega_b + {\Omega_b}^2 + 1 - 3\zeta_c \tag{21}$$

and Ω_b is the smallest positive root of the following equation:

$$\Omega_b{}^3 + (2 - 3\zeta_c)\Omega_b{}^2 + 3\zeta_c{}^2\Omega_b - \zeta_c{}^3 = 0 \tag{22}$$

For $\alpha[T_r]$, the same function of reduced temperature that was described for the SRK and PR was used (see Eq. (8)).

$$\zeta_c = \frac{P_c V_c}{RT_c} \tag{23}$$

The critical compressibility factor ζ_c is treated as an empirical parameter. The optimal values of ζ_c and m correspond to the minimum deviation in saturated liquid densities and the equilibrium condition of equality of fugacities. Therefore, the proposed equation using optimum values of ζ_c and m can provide good prediction of liquid phase densities and vapor-liquid equilibria [43]. For nonpolar substances, ζ_c and m are calculated as follows:

$$\zeta_c = 0.329 - 0.077\omega + 0.021\omega^2 \tag{24}$$

$$m = 0.452 + 1.310\omega - 0.296\omega^2 \tag{25}$$

4.2 Mixing rules

In order to adapt the above EoS to a binary mixture, the characteristics parameters a, b, and c must be obtained using mixing rules. Commonly, the "one fluid" wdW mixing rules are used when the mixture is assumed to behave as a pure component with appropriate parameters. It can be expressed as follows:

$$a_{mix} = \sum_i \sum_j x_i x_j a_{ij} \tag{26}$$

$$a_{ij} = \sqrt{a_i a_j}\left(1 - k_{ij}\right) \tag{27}$$

where a_{ij} is a cross-parameter and k_{ij} is the binary interaction parameter obtained from the correlation of the experimental data.

$$b = \sum_i x_i b_i \tag{28}$$

$$c = \sum_i x_i c_i \tag{29}$$

For normal fluids, the above equations can give reasonable correlation with the experimental data. However, modifications of the wdW mixing rules for polar or highly non-ideal compounds are required. For instance, Panagiotopoulos and Reid [44] defined a two-parameter mixing rule, for the a_{ij} term and another binary interaction parameter k_{ji} (Eq. (30)):

$$a_{ij} = \sqrt{a_i a_j}\left(1 - k_{ij} + \left(k_{ij} - k_{ji}\right)x_i\right) \tag{30}$$

The Panagiotopoulos and Reid mixing rule and others [42, 45] allow that the EoS would be adapted at different compositions. More recently, Mathias et al. [46] proposed the well-known Mathias-Klotz-Prausnitz (MKP) mixing rule that is shown below:

$$a = \sum_i \sum_j x_i x_j \sqrt{a_i a_j}(1 - k_{ij}) + \sum_i x_i \left(\sum_j x_j \left(\sqrt{a_i a_j} \right)^{\frac{1}{3}} l_{ji}^{\frac{1}{3}} \right)^3 \tag{31}$$

where k_{ij} is a symmetric binary parameter and l_{ji} is antisymmetric.

With this approach, the MKP mixing rules has contributed to a better representation of the phase behavior of highly non-ideal mixtures and systems formed by two very similar components and a very different third one [46].

5. Conclusion

From the review of experimental methods and thermodynamic models to describe the behavior of high-pressure systems, it can be concluded that the applicability of SCF depends largely on the availability of models that allow us to predict phase equilibria in the systems involved. There is an extensive amount of experimental data on high-pressure phase equilibria in the literature, although SCF systems are particularly difficult due to the high compressibility and asymmetry of most systems of interest.

A large number of experimental methods have been developed for the measurement of the phase equilibria of fluids at high pressure, both analytical, involving the analytical determination of the compositions of the co-existing phases, and synthetic, which are based on the preparation of a mixture of precisely known composition, the observation of phase behavior in an equilibrium cell, and the measurement of properties in the equilibrium state.

For high-pressure binary systems, a wide variety of phase behaviors can occur. Van Konynenburg and Scott's classification (six types of pressure-temperature diagrams, considering the van der Waals equation of state and the rules of quadratic mixture) have been used to analyze the phase behavior of binary mixtures.

An equation of state (EoS) is an algebraic relation between P, V, and T, which serves for describing some behavior of Nature. The most common EoS models have been described in this chapter. The cubic wdW EoS, based on molecular interactions of attraction and repulsion, can quantitatively predict most of the phase equilibrium behaviors exhibited by binary mixtures. Mixing rules are necessary to adapt EoS to binary mixtures.

DOI: http://dx.doi.org/10.5772/intechopen.105486

Author details

Mercedes G. Montalbán* and Gloria Víllora
Faculty of Chemistry, Chemical Engineering Department, Regional Campus of International Excellence "Campus Mare Nostrum", University of Murcia, Murcia, Spain

*Address all correspondence to: mercedes.garcia@um.es

References

[1] Fornari RE, Alessi P, Kikic I. High pressure fluid phase equilibria: Experimental methods and systems investigated (1978–1987). Fluid Phase Equilibria. 1990;**57**(1):1-33

[2] Dohrn R, Brunner G. High-pressure fluid-phase equilibria: Experimental methods and systems investigated (1988–1993). Fluid Phase Equilibria. 1995;**106**(1):213-282

[3] Christov M, Dohrn R. High-pressure fluid phase equilibria: Experimental methods and systems investigated (1994–1999). Fluid Phase Equilibria. 2002;**202**(1):153-218

[4] Dohrn R, Peper S, Fonseca JMS. High-pressure fluid-phase equilibria: Experimental methods and systems investigated (2000–2004). Fluid Phase Equilibria. 2010;**288**(1–2):1-54

[5] Fonseca JMS, Dohrn R, Peper S. High-pressure fluid-phase equilibria: Experimental methods and systems investigated (2005–2008). Fluid Phase Equilibria. 2011;**300**(1–2):1-69

[6] Peper S, Fonseca JMS, Dohrn R. High-pressure fluid-phase equilibria: Trends, recent developments, and systems investigated (2009–2012). Fluid Phase Equilibria. 2019;**484**:126-224

[7] Thamanavat K. High Pressure Phase Equilibria in the Carbon Dioxide + Pyrrole System [Master Thesis]. USA: Georgia Institute of Technology; 2004

[8] Muhlbauer AL, Raal J. Measurement and thermodynamic interpretation of high-pressure vapour_liquid equilibria in the tolueneCO2 system. Fluid Phase Equilibria. 1991;**64**:213-236

[9] Nakayama T, Sagara H, Arai K, Saito S. High pressure liquid-liquid equilibria for the system of water, ethanol and 1,1-difluoroethane at 323.2 K. Fluid Phase Equilibria. 1987;**38**(1): 109-127

[10] Zimmermann A, Keller JU. VLE In the system water-ammonia-lithium bromide. Fluid Phase Equilibria. 1989;**53**: 229-234

[11] Ho QN, Yoo KS, Lee BG, Lim JS. Measurement of vapor–liquid equilibria for the binary mixture of propylene (R-1270)+propane (R-290). Fluid Phase Equilibria. 2006;**245**(1):63-70

[12] Kodama D, Miyazaki J, Kato M, Sako T. High pressure phase equilibrium for ethylene + 1-propanol system at 283.65K. Fluid Phase Equilibria. 2004; **219**(1):19-23

[13] Beier A, Kuranov J, Stephan K, Hasse H. High-pressure phase equilibria of carbon dioxide + 1-Hexanol at 303.15 and 313.15 K. Journal of Chemical & Engineering Data. 2003;**48**(6):1365-1367

[14] Fang T, Shimoyama Y, Abeta T, Iwai Y, Sasaki M, Goto M. Phase equilibria for the mixtures of supercritical methanol+C18 methyl esters and supercritical methanol+α-tocopherol. Journal of Supercritical Fluids. 2008;**47**(2):140-146

[15] Shimoyama Y, Iwai Y, Jin BS, Hirayama T, Arai Y. Measurement and correlation of vapor–liquid equilibria for methanol+methyl laurate and methanol +methyl myristate systems near critical temperature of methanol. Fluid Phase Equilibria. 2007;**257**(2):217-222

[16] Fang T, Wahyudiono, Al-Duri B, Shimoyama Y, Iwai Y, Sasaki M, et al. Supercritical methanol process of modifying oil byproduct for

concentrating natural tocopherols. Industrial and Engineering Chemistry Research. 2007;**46**(16):5325-5332

[17] Sawada K, Oshima M, Sugimito M, Ueda M. Microemulsions in supercritical CO2 utilizing the polyethyleneglycol dialkylglycerol and their use for the solubilization of hydrophiles. Dyes and Pigments. 2005;**65**(1):67-74

[18] Jödecke M, Pérez-Salado Kamps Á, Maurer G. Experimental investigation of the solubility of CO2 in (acetone + water). Journal of Chemical & Engineering Data. 2007;**52**(3): 1003-1009

[19] Seo Y, Lee S, Lee H. Experimental measurements of hydrate phase equilibria for carbon dioxide in the presence of THF, propylene oxide, and 1,4-dioxane. Journal of Chemical & Engineering Data. 2008;**53**(12): 2833-2837

[20] Mohammadi AH, Afzal W, Richon D. Experimental data and predictions of dissociation conditions for ethane and propane simple hydrates in the presence of methanol, ethylene glycol, and triethylene glycol aqueous solutions. Journal of Chemical & Engineering Data. 2008;**53**(3):683-686

[21] Carissimi G, Montalbán MG, Díaz Baños FG, Víllora G. High pressure phase equilibria for binary mixtures of CO_2 + 2-pentanol, vinyl butyrate, 2-pentyl butyrate or butyric acid systems. Journal of Supercritical Fluids. 2018;**135**: 69-77

[22] Thamanavat K, Sun T, Teja AS. High-pressure phase equilibria in the carbon dioxide + pyrrole system. Fluid Phase Equilibria. 2009;**275**(1):60-63

[23] Fonseca JMS, von Solms N. Development and testing of a new apparatus for the measurement of high-pressure low-temperature phase equilibria. Fluid Phase Equilibria. 2012; **329**:55-62

[24] Taib MM, Murugesan T. Solubilities of CO2 in aqueous solutions of ionic liquids (ILs) and monoethanolamine (MEA) at pressures from 100 to 1600kPa. Chemical Engineering Journal. 2012;**181–182**:56-62

[25] Gui X, Tang Z, Fei W. CO2 capture with physical solvent dimethyl carbonate at high pressures. Journal of Chemical & Engineering Data. 2010;**55**(9):3736-3741

[26] Ruiz F, Marcilla A, Ancheta AM, Rico C. Liquid-liquid equilibrium of the three liquid phases at equilibrium system water-phosfhoric acid-di-isopropyl ether at 25 and 40°C. Solvent Extraction and Ion Exchange. 1986;**4**(4):789-802

[27] van Konynenburg PH, Scott RL, Rowlinson JS. Critical lines and phase equilibria in binary van der Waals mixtures. Philosophical Transactions of the Royal Society London Series A, Mathematical, Physical and Engineering Sciences. 1980;**298**(1442):495-540

[28] Coquelet C, Ramjugernath D. Phase diagrams in chemical engineering: Application to distillation and solvent extraction. In: Nawaz Z, Naveed S, editors. Advances in Chemical Engineering. Rijeka: IntechOpen; 2012

[29] McHugh MA, Krukonis VJ. In: Brenner H, editor. Supercritical Fluid Extraction: Principles and Practice. 2nd ed. Stoneham, MA, USA: Butterworth-Heinemann; 1994. p. 608

[30] Wei MS-W, Brown TS, Kidnay AJ, Sloan ED. Vapor + liquid equilibria for the ternary system methane + ethane + carbon dioxide at 230 K and its constituent binaries at temperatures

from 207 to 270 K. Journal of Chemical & Engineering Data. 1995;**40**(4):726-731

[31] Chen H-I, Chang H-Y, Chen P-H. High-pressure phase equilibria of carbon dioxide + 1-butanol, and carbon dioxide + water + 1-butanol systems. Journal of Chemical & Engineering Data. 2002; **47**(4):776-780

[32] Lam DH, Jangkamolkulchai A, Luks KD. Liquid—Liquid—Vapor phase equilibrium behavior of certain binary carbon dioxide + n-alkanol mixtures. Fluid Phase Equilibria. 1990;**60**(1): 131-141

[33] Sima S, Ioniță S, Secuianu C, Feroiu V, Geană D. High-pressure phase equilibria of carbon dioxide + 1-octanol binary system. Journal of Chemical & Engineering Data. 2018;**63**(4):1109-1122

[34] Privat R, Jaubert J-N. Classification of global fluid-phase equilibrium behaviors in binary systems. Chemical Engineering Research and Design. 2013; **91**(10):1807-1839

[35] Kodama D, Kimura T, Tanaka H, Kato M. High pressure phase equilibrium properties for ethane+1-butanol system at 313.15 K. Fluid Phase Equilibria. 2002; **201**(2):401-407

[36] Hou Y, Chen X, Ren S, Li G, Wu W. High pressure phase behavior of carbon dioxide+nitrobenzene binary system at 298.15, 310.45 and 322.75K. Journal of Supercritical Fluids. 2009;**51**(2):104-108

[37] Poling BE, Praustniz JM, O'Connell JP. The Properties of Gases and Liquids. 5th ed. New York, USA: McGraw-Hill; 2001. p. 803

[38] van der Waals JD, Rowlinson JS. On the Continuity of the Gaseous and the Liquid States. 1st ed. Mineola, New York: Dover Publications; 2004

[39] Scott RL, van Konynenburg PH. Static properties of solutions. Van der Waals and related models for hydrocarbon mixtures. Discussions of the Faraday Society. 1970;**49**(0):87-97

[40] Peng D-Y, Robinson DB. A new two-constant equation of state. Industrial and Engineering Chemistry Fundamentals. 1976;**15**(1):59-64

[41] Soave G. Equilibrium constants from a modified Redlich-Kwong equation of state. Chemical Engineering Science. 1972;**27**(6):1197-1203

[42] Stryjek R, Vera JH. PRSV: An improved peng—Robinson equation of state for pure compounds and mixtures. Canadian Journal of Chemical Engineering. 1986;**64**(2):323-333

[43] Patel NC, Teja AS. A new cubic equation of state for fluids and fluid mixtures. Chemical Engineering Science. 1982;**37**(3):463-473

[44] Panagiotopoulos AZ, Reid RC. New Mixing Rule for Cubic Equations of State for Highly Polar, Asymmetric Systems. Vol. 300. USA: American Chemical Society (ACS Symposium Series); 1986. pp. 571-582

[45] Adachi Y, Sugie H. A new mixing rule—Modified conventional mixing rule. Fluid Phase Equilibria. 1986;**28**(2): 103-118

[46] Mathias PM, Klotz HC, Prausnitz JM. Equation-of-state mixing rules for multicomponent mixtures: The problem of invariance. Fluid Phase Equilibria. 1991;**67**:31-44

Chapter 3

Application of Supercritical Phase Equilibria to the Components of the Transesterification Reaction of *rac*-2-Pentanol with a Vinyl Ester

Mercedes G. Montalbán and Gloria Víllora

Abstract

This chapter illustrates the collection of phase equilibrium and high-pressure solubility data applied to four binary systems, (CO_2 + 2-pentanol, CO_2 + vinyl butyrate, CO_2 + 2-pentyl butyrate and CO_2 + butyric acid) at three temperatures of (313.15, 323.15, and 333.15) K and pressures up to 11 MPa. These four organic compounds were selected because they are implicated in the kinetic resolution of *rac*-2-pentanol, and their phase equilibria play an important role in the separation processes of the reaction compounds. Equilibrium data were obtained using a synthetic method in a high-pressure cell of variable volume. All systems were found to have type I phase behavior. Experimental high-pressure data showed a good correlation with density-based models and by the well-known Peng-Robinson (PR) and Soave-Redlich-Kwong (SRK) EoS coupled with the quadratic mixing rule in a semipredictive approach to describe the phase equilibrium topology of the four binary mixtures.

Keywords: high-pressure phase equilibrium, supercritical carbon dioxide, equation of state, mixing rule, 2-pentanol, vinyl butyrate, 2-pentyl butyrate, butyric acid

1. Introduction

The chapter presents a work based on the study of the behavior in supercritical medium of the reactants and products of the transesterification reaction of *rac*-2-pentanol with vinyl butyrate. A particular feature of this reaction is that *rac*-2-pentanol contains an asymmetrical carbon, and hence, it is formed by a racemic mixture of two enantiomers (R and S). The lipase enzyme that catalyzes this reaction is stereoselective so it favors the reaction of only one of the enantiomeric forms of the *rac*-2-pentanol, specifically the (R)-2-pentanol.

This reaction is of great importance from the point of view of the pharmaceutical industry because (S)-2-pentanol is obtained, and this is a basic intermediate compound in the synthesis of drugs against the Alzheimer disease. In general, the reaction of *rac*-2-pentanol with any vinyl ester is represented in **Figure 1**. This figure shows the reactants, vinyl ester (R-COO-CH=CH_2) and *rac*-2-pentanol, the compounds

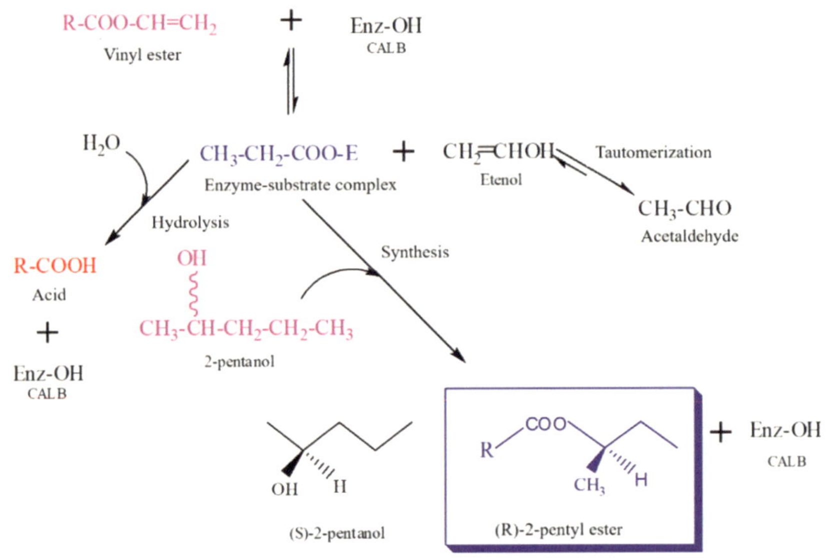

Figure 1.
Stoichiometric scheme of racemic resolution of rac-2-pentanol catalyzed by a lipase.

obtained by the synthesis route, (R)-2-pentyl ester and (S)-2-pentanol, and the compound that results from the hydrolysis reaction, the acid. The hydrolysis is the undesired reaction and parallelly competes with the desired reaction.

As can be seen, the first step consists of the formation of an enzyme-substrate complex between the vinyl ester and the lipase enzyme. As a consequence of the formation of this intermediate, vinyl alcohol is released, which by presenting a hydrogen atom attached to one of the carbon atoms adjacent to the carbonyl group is forming a keto-enolic equilibrium highly displaced toward the keto form, i.e., it is found as acetaldehyde. This fact contributes to removing vinyl alcohol from the medium and preventing that the vinyl ester will be formed again (which would be the reverse reaction to the desired reaction), being possible to achieve large final conversions, by favoring the displacement of equilibrium.

In the second reaction stage, there can be two possibilities:

- Hydrolysis: it is the reaction of the enzyme-substrate complex with water, giving rise to an acid and the regeneration of the enzyme in its initial state.

- Synthesis: the reaction of the complex with *rac*-2-pentanol forming the (R)-2-pentyl ester, also regenerating the enzyme.

As can be deduced from the above, the water present in the medium plays a fundamental role in the formation of (R)-2-pentyl ester, which is the product of interest. As the kinetic constant of hydrolysis is much greater than that of synthesis, for synthesis to predominate over hydrolysis, the concentration of the second substrate must be much higher than that of water, since this will achieve that:

$$k_H \cdot [H_2O] << k_S \cdot [rac\text{-}2\text{-pentanol}]$$

where $[H_2O]$ is the concentration of water in the medium, [*rac*-2-pentanol] is the concentration of the second substrate, and k_H and k_S are the hydrolysis and synthesis constants, respectively. Accordingly, if the amount of water in the medium is controlled, it will be possible to minimize the speed of the hydrolysis reaction compared with that of synthesis.

Although an overview has been given here, in this particular work the vinyl ester studied will be vinyl butyrate, and the products are those corresponding to this ester: 2-pentyl butyrate and butyric acid.

The study of phase equilibria of mixtures involving CO_2 and other compounds at high pressure and under supercritical conditions is of crucial importance for their application in a wide range of operations related to the chemical industry such as reaction, extraction, fractionation, separation of mixtures, supercritical chromatography, synthesis, and/or fabrication of nanostructured porous materials metal support, or formation of nanocrystals, etc. [1]. It is therefore essential that experimental equilibrium data are reliable and accurate for the optimization of the involved processes [2].

Due to the high number of applications of $scCO_2$, many authors have studied in depth the behavior of systems (CO_2 + organic compounds) under high-pressure conditions. Isothermal data from binary systems CO_2+ different alcohols have been measured and correlations have been obtained for prediction purposes [1, 3–10]. Thermodynamic knowledge of the high-pressure phase behavior of carbon dioxide + alcohol mixtures is essential for the design and optimization of many supercritical fluid extraction and supercritical fluid chromatography processes in sectors of great importance such as the oil and natural gas industry and in the food, pharmaceutical, cosmetic, and surfactant industries [1]. High-pressure equilibrium data from other organic compounds with $scCO_2$ can be found in the literature, such as alkanes [11], acids [12], ketones [13], amides [13], aromatic compounds such as pyrrole [14, 15], furans [16], or nitriles [17, 18]. As stated in Chapter 2, in recent decades, several reviews have been published that collect a large number of experimental data of high-pressure phase equilibrium in different systems, most of which are binary systems involving CO_2 [19–21]. These reviews classify the results based on the experimental procedure used to obtain them.

As an example, this chapter shows how to obtain phase equilibrium and high-pressure solubility data applied to four binary systems (CO_2 + organic compound). The four organic components are those involved in the kinetic resolution of *rac*-2-pentanol by transesterification of vinyl butyrate catalyzed by a lipase. The scheme in **Figure 1** shows the stoichiometry of the reaction. The compound of interest is (*S*)-2-pentanol, which is a key intermediate necessary for the synthesis of several drugs for the treatment of Alzheimer's disease that inhibit the release and/or synthesis of β-amyloid peptide [22]. The enzymatic resolution of *rac*-2-pentanol by *Candida antarctica* lipase B has been demonstrated. Commercially available *C. antarctica* lipase B efficiently catalyzed the enantioselective acetylation of *rac*-2-pentanol yielding an enantiomeric excess (ee) of 99% for (*S*)-2-pentanol [23].

For long, the excellent properties that $scCO_2$ has for the dissolution, extraction, and transport of chemical compounds are well known. This is due to its low viscosity, low surface tension, and high diffusion coefficients. The diffusivity of the substrates in $scCO_2$ allows an excellent mass transfer and is a clean alternative to conventional organic solvents that can contribute to the integration of reaction and separation processes in a single stage. The measurement of the high-pressure phase equilibrium data of binary mixtures (CO_2 + 2-pentanol, vinyl butyrate, 2-pentyl butyrate, or butyric acid) and the determination of the solubility between CO_2 and the organic compound would make it possible to establish the ability of CO_2 to separate reaction products and optimize the operating conditions to carry out such separation. Nevertheless,

$scCO_2$ could have an adverse effect on the enzyme and cause deactivation due to the decrease in the pH of the enzyme microenvironment, to the covalent modification of the free amino groups on the surface of the protein forming carbamates and/or to the pressurization/depressurization cycles [24]. As discussed in Chapter 2, to solve these possible problems, the use of $scCO_2$/ionic liquid biphasic systems has been proposed. Ionic liquids have also been considered "green" solvents due to their negligible vapor pressure, and they can contribute to supply of a nonaqueous catalytic medium [25]. The success of these biphasic systems is based on the practical insolubility of the ionic liquid in $scCO_2$ and the high solubility of $scCO_2$ in the ionic liquid, in such a way that the extraction of organic compounds from the ionic liquid is facilitated by $scCO_2$ without cross-contamination of the organic compound with the ionic liquid [26, 27].

One of the objectives of this study was to obtain experimental measurements of the high-pressure phase equilibrium of the systems (CO_2 + 2-pentanol), (CO_2 + vinyl butyrate), (CO_2 + 2-pentyl butyrate), and (CO_2 + butyric acid) in isothermal tests at temperatures of 313.15, 323.15, and 333.15 K. Another aim was to obtain the correlations of the experimental data using density-based models and by Peng-Robinson (PR) [28] and Soave-Redlich-Kwong (SRK) [29] EoS together with the quadratic mixture rule providing a semipredictive approach that describes the phase equilibria of the four binary systems. To obtain the correlations, the acentric factor (ω) was estimated using the Lee-Kessler group contribution method, and the critical temperature and critical pressure (T_c, P_c,) values of the organic compounds were obtained from the literature or were estimated using the Joback group contribution method.

2. Apparatus and experimental procedure

To obtain the experimental values of phase equilibrium of binary systems (CO_2 + 2-pentanol), (CO_2 + vinyl butyrate), (CO_2 + 2-pentyl butyrate), and (CO_2 + butyric acid), the commercial Super Phase Monitor equipment (SPM system, Thar Technologies, Inc., USA) was used, which is shown in the scheme of **Figure 2**. The most important elements of the device are: a high-pressure pump, a thermostated cell of variable volume, provided with two sapphire windows to locate a camera and a lighting source, a visual device, and a control software provided by the supplier. The cell has an agitator and a piston cylinder that moves by a hydraulic pump. The volume of the cell is determined by the position of the piston. The CO_2 is fed to the cell by a syringe pump (Teledyne ISCO, model 500D, USA), provided with a pressure controller. The pump has a cooling system connected to a thermostatic water bath (Frigiterm, J.P. Selecta S.A., Spain) that keeps the CO_2 in liquid phase. The measuring cell allows visual observation of its interior through the sapphire windows, and the image is collected in a video system that displays the image on the computer monitor.

A synthetic method was used to carry out the measurements of phase equilibria. The procedure followed has been described elsewhere [30]. Briefly, first, the cell was purged with CO_2 at low pressure to remove the air from inside of the cell. The organic compound was then weighed with an accuracy of 0.1 mg with an analytical scale (Sartorius, model ED 1245, Germany) and loaded into the cell with a syringe. Then compressed CO_2 was introduced into the cell using the syringe pump. The molar fraction of the binary mixture within the cell was determined on the base of the CO_2 loaded mass, calculated by the volume displaced by the pump and the CO_2 density, obtained from NIST [31], at the selected pressure and temperature values.

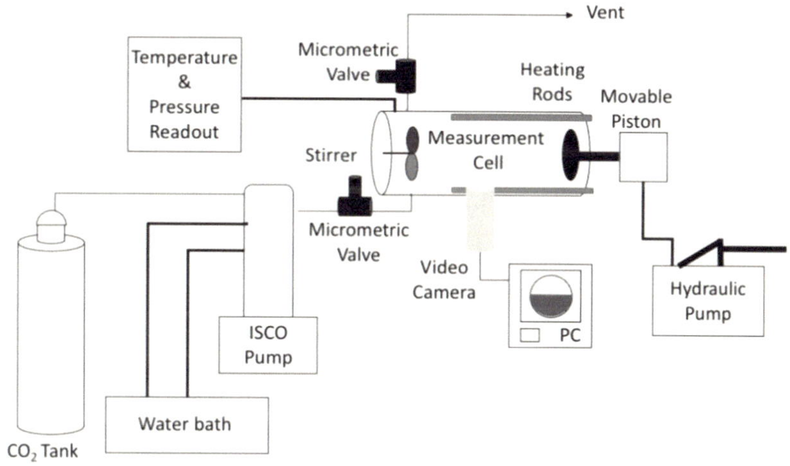

Figure 2.
Scheme of the high-pressure phase equilibrium apparatus. From reference [30] with permission of Elsevier.

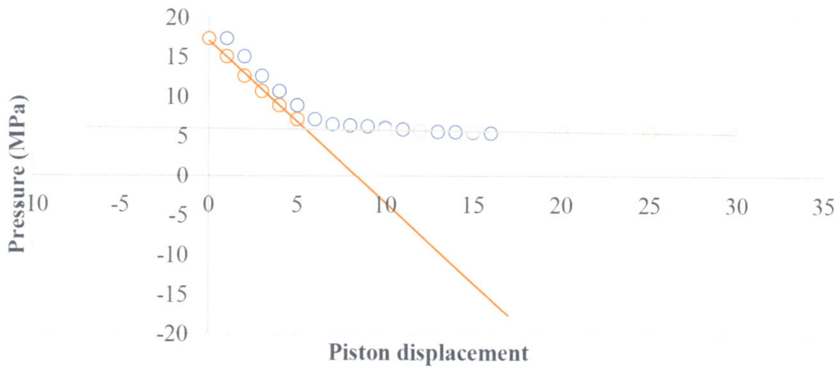

Figure 3.
Pressure-volume relation for the depressurization of (CO_2 + 2-pentanol) at 313.15 K; CO_2 mole fraction: 0.541.

It was estimated that the mass of CO_2 when each experiment begins, after purging, was <1% and was ignored.

With the components inside the cell, the temperature was set at the selected value and the pressure was raised by moving the plunger to reduce the volume of the cell until a homogeneous phase was observed. The system was then kept in agitation for about 30 minutes to ensure that the system was in equilibrium. From this moment, the volume was increased by 80 μL in a staggered way, displacing the piston, and recording the pressure of the cell until a second phase was observed. Each experiment was repeated at least twice. **Figure 3** shows, as an example, the determination of the bubble point from a graph of depressurization of a mixture (CO2 + 2-pentanol) at 313.15 K (molar fraction of CO_2: 0.541). As can be observed, the pressure displacement of the piston (correlated with cell volume) line changes slope when a new phase with a different isothermal compressibility appears. The intersection of the two lines corresponds to the equilibrium pressure, as described by Thamanavat et al. [14, 32]. This procedure was repeated at temperatures of 313.15, 323.15, and 333.15 K to obtain isothermal curves.

3. Results and discussion

3.1 Experimental phase equilibrium data

Figures 4 and **5** show the comparison of the experimental values and those calculated from Peng-Robinson and Soave-Redlich-Kwong EoS and quadratic mixing rule for the systems studied in the form of isotherms (313.15, 323.15, and 333.15 K) of pressure composition (P, x). These figures also show the bubble points measured for all systems. With respect to the experimental measurements previously published [2, 9, 33, 34], the data reported by the authors agree, with deviations less than 0.3 MPa, even they are even more accurate due to the modifications made in the experimental procedure. As can be seen in **Figures 5** and **6**, the bubble pressure increases when the molar fraction of CO_2 is increased for all isotherms, indicating that the liquid phase, rich in organic compound, can dissolve more CO_2. However, at a constant pressure, the solubility of CO_2 in the above phase is reduced for all systems when the temperature is increased. In this work, for the temperature and molar fraction values used, three phases were not detected, but all binary mixtures exhibited critical mixture curves with a maximum pressure-temperature located in the range of critical temperatures of the organic compound and CO_2. This phase behavior corresponds in all cases to Type I according to the Scott and Konynenburg classification [35]. As explained in Chapter 2, it is represented by a critical curve that links up the critical points of pure compounds in the P-T phase diagram.

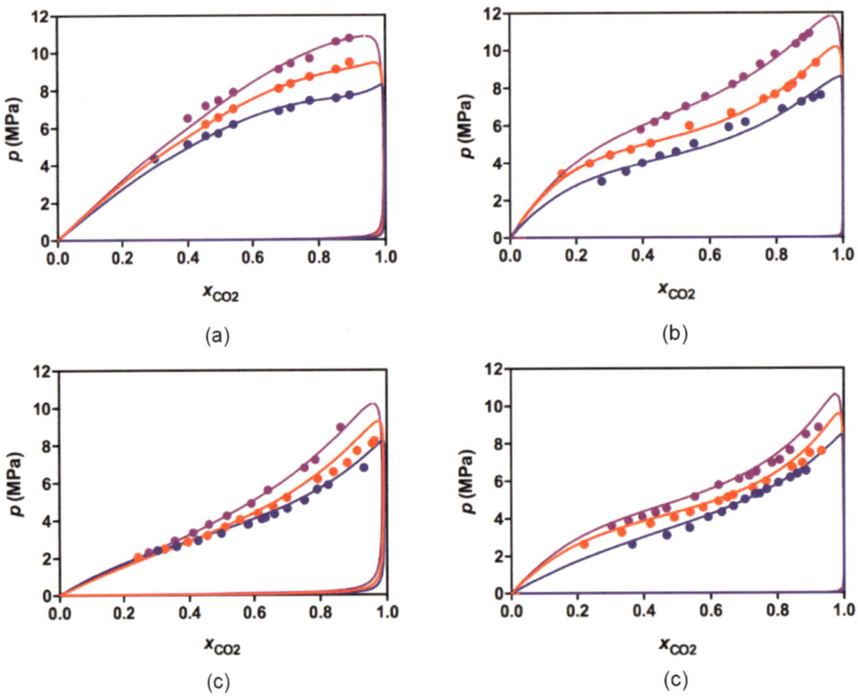

Figure 4.
Experimental data and calculated phase equilibrium behavior for carbon dioxide + organic compound at ● 313.15 K, ● 323.15 K, and ● 333.15 K with Peng-Robinson EoS and quadratic mixing rule. (a) 2-pentanol; (b) butyric acid; (c) vinyl butyrate; (d) 2-pentyl butyrate. From reference [30] with permission of Elsevier.

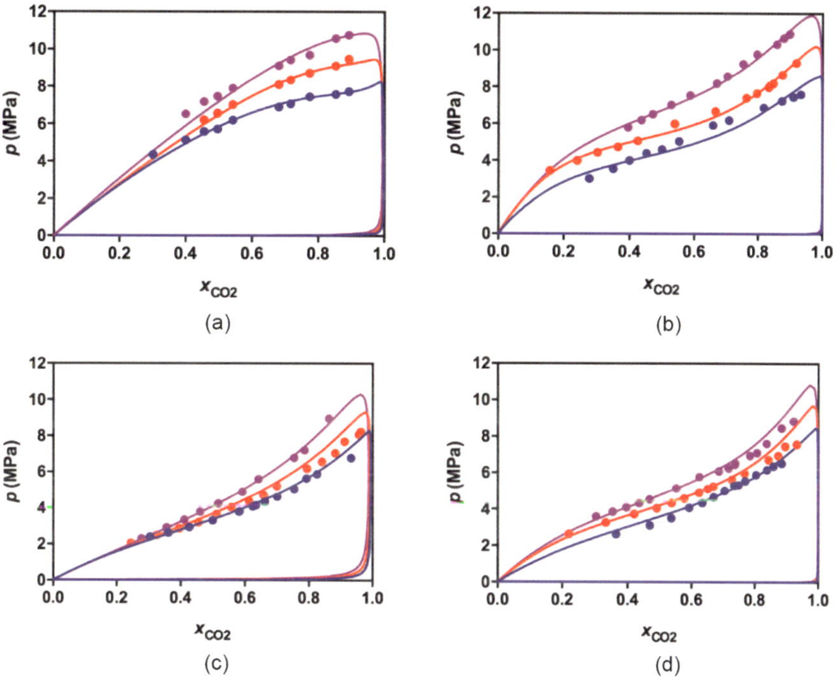

Figure 5.
Experimental data and calculated phase equilibrium behavior for carbon dioxide + organic compound at ● 313.15 K, ● 323.15 K, and ● 333.15 K with Soave-Redlich-Kwong EoS and the quadratic mixing rule. (a) 2-pentanol; (b) butyric acid; (c) vinyl butyrate; (d) 2-pentyl butyrate. From reference [30] with permission of Elsevier.

3.2 Density-based models

The density of CO_2 was obtained by the Chrastil Equation [36], which is a linear equation that relates the solubility of the organic compound and the density of $scCO_2$. This equation has been used to study supercritical gaseous solutions with low solute concentration and to determine the accuracy of experimentally obtained solubility measurements.

Henry's law is normally used to study the solubility of a gas in a liquid. It shows that the amount of gas that can be dissolved in a liquid depends linearly on its partial pressure. However, some deviations from Henry's law are observed at high pressures and, then, it is not suitable for obtaining the solubility of supercritical fluids in organic liquids. The semiempirical Eqs. (1) and (2) were obtained, modifying Henry's law, by Hernández et al. [37] to determine the solubility of a supercritical fluid in a liquid in terms of the molar fraction of the supercritical fluid in the liquid phase. Eq. (1) gives better solubility values when positive deviations to Henry's law are presented for the mixture, and Eq. (2) is more suitable for mixtures with negative deviations from Henry's law.

$$\ln\left(x_{scCO2}\right) = A\ln\left(\rho_{scCO2}\right) + B\left(constanttemperature\right) \tag{1}$$

$$\ln\left(x_{scCO2}\right) = A'\frac{\ln\left(\rho_{scCO2}\right)}{P} + B'\left(constanttemperature\right) \tag{2}$$

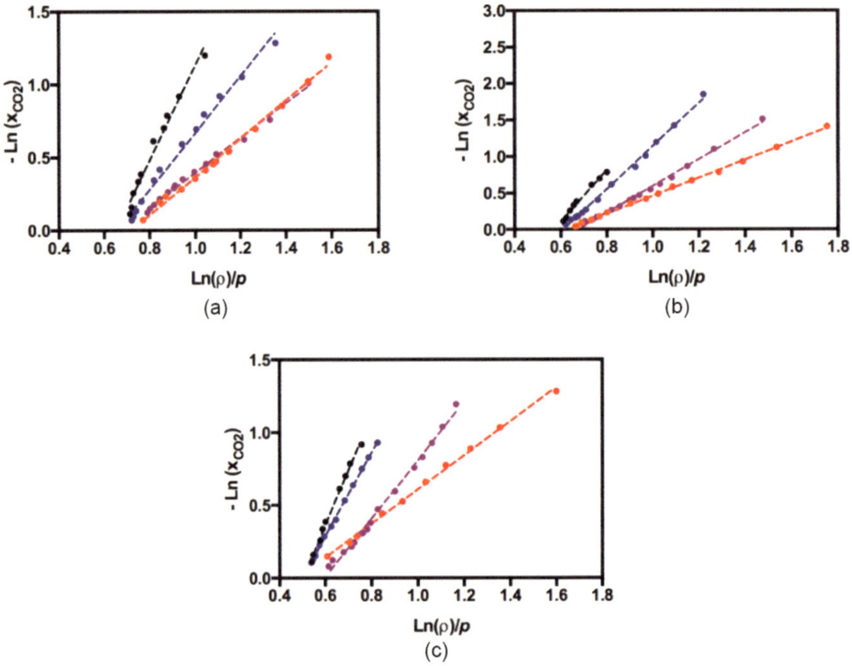

Figure 6.
Linear correlation between ln(x_{CO_2}) and ln(ρ_{CO_2}/p) for binary mixtures of CO_2 with 2-pentanol (●), butyric acid (●), vinyl butyrate (●), and 2-pentyl butyrate (●) at a) 313.15 K; b) 323.15 K; c) 333.15 K. from reference [30] with permission of Elsevier.

where x_{scCO2} is the mole fraction of $scCO_2$, P is the pressure (MPa), ρ_{scCO2} *is* the supercritical carbon dioxide density (kg/m^3), and A, B, A', and B' are the fitting parameters, which are obtained from the regression of the experimental data. **Table 1** shows the fitting parameters and the regression coefficients, R^2 and $R^{2'}$, obtained for the studied binary systems.

Figure 6 shows the results of the fittings at 313.15, 323.15, and 333.15 K. The following order was found for the solubility of CO_2 in the organic compounds for all temperatures: 2-pentanol > butyric acid > esters. From these results, it can be induced that the relative polarity of CO_2 and the organic compounds has an important contribution to the CO_2 solubility [38]. From these results, it can be concluded that the unreacted (S)-2-pentanol could be recovered from a biphasic system liquid ionic/CO_2, since the solubility of CO_2 in 2-pentanol is higher than that in the other compounds involved in the transesterification reaction and would allow the successful separation of the racemic mixture.

3.3 Thermodynamic modeling

As was reported on Chapter 2, the available literature shows a vast compilation of EoS to predict phase behavior of binary systems. Among them, Peng-Robinson [28] and Soave-Redlich-Kwong [29] EoS, which are cubic equations, are widely used because of their simplicity and accuracy. The experimental data obtained in this study were fitted using the two EoS mentioned above, which were combined with quadratic (Q) mixing rules with two binary interaction parameters (BIP) designated as k_{ij}, l_{ij}.

Compound	T (K)	CO_2 density range ($kg.m^{-3}$)	A	-B	R^2	-A'	B'	$R^{2'}$
2-Pentanol	313.15	93.4–250.7	1.0647	5.9939	0.9949	3.2701	2.1380	0.9750
	323.15	142.0–327.7	0.8269	4.8724	0.9965	3.5596	2.0295	0.9864
	333.15	140.3–339.6	0.9068	5.3782	0.9929	3.7674	1.9017	0.9940
Butyric acid	313.15	59.1–238.7	0.8498	4.6715	0.9895	1.9381	1.2714	0.9876
	323.15	65.5–312.4	1.1091	6.1856	0.9334	2.9343	1.7907	0.9958
	333.15	118.8–349.0	0.7498	4.4433	0.9751	2.8796	1.4424	0.9989
Vinyl butyrate	313.15	45.7–185.1	0.7676	3.9554	0.9393	1.3095	0.9444	0.9944
	323.15	36.5–232.2	0.6835	3.6511	0.9504	1.2351	0.7730	0.9991
	333.15	39.9–233.1	0.6537	3.5833	0.9651	1.1726	0.5646	0.9976
2-Pentyl butyrate	313.15	50.1–172.4	0.6571	3.4738	0.9746	1.1812	0.7841	0.9945
	323.15	48.2–197.7	0.9353	4.8913	0.9328	1.8270	1.2287	0.9953
	333.15	65.8–227.8	0.8914	4.7735	0.9474	2.0089	1.2002	0.9934

Table 1.
Fitting parameters of Eqs. (1) and (2) for (CO_2 + 2-pentanol, butyric acid, vinyl butyrate, or 2-pentyl butyrate) binary systems at 313.15, 323.15, and 333.15 K. from reference [30] with permission of Elsevier.

To correlate the data, the Phase Equilibria 2000 (PE2000) software, developed by Brunner et al. [39], was used. This software employs the Simplex-Nelder-Mead algorithm [40] to optimize binary interaction parameters. Critical parameters were estimated by the contribution method of the Joback group [41] or taken from the literature [42–45], and acentric factors were estimated by the Lee-Kessler group contribution method [46]. **Table 2** collects the values of the critical parameters used.

Figures 4 and **5** show the experimental and calculated data obtained with EoS. A great agreement was found between the experimental and calculated data for all mixtures with Peng-Robinson and Soave-Redlich-Kwong EoS.

Table 3 shows the BIP used with the Peng-Robinson and Soave-Redlich-Kwong EoS for each system and temperature. The suitability of using the modeling approach to correlate the experimental bubble point data with the points calculated by Peng-Robinson and Soave-Redlich-Kwong EoS was assessed from the relative mean square deviation (RMSD). The found values are collected in **Table 4**.

As a consequence of the analysis of the results, it can be observed that, for pressures close to the critical point of the system, the molar fraction of the organic component in the vapor phase increased for all the mixtures studied as the temperature increased at constant pressure. This means that the solvation capacity of the vapor phase (rich in CO_2) increases when its density increased to reach the critical point of the mixture. In addition, when the temperature increases, the chemical potential of the organic compound also increases.

The experimental results and the results obtained from the density-based and thermodynamic models allow us to infer that the different solubility of organic compounds in CO_2 could be useful to carry out the kinetic resolution of the racemic mixture of (*R*, *S*)-2-pentanol at different temperature values. In the racemic resolution of *rac*-2-pentanol, the (*S*)-2-pentanol does not react with the vinyl ester and can be dissolved in $scCO_2$. However, (*R*)-2-pentanol reacts with vinyl butyrate when an

Compound	Mw (g/mol)	p_c (MPa)	T_c (K)	ω
Carbon Dioxide	44.10	7.38[a]	304.2[a]	0.225[a]
Acetone	58.08	4.70[b]	508.1[b]	0.306[**]
2-Pentanol	88.20	3.68[c]	560.3[c]	0.557[**]
Butyric Acid	88.11	4.06[d]	615.2[d]	0.690[**]
Vinyl Butyrate	114.14	3.56[*]	539.9[*]	0.394[**]
2-Pentyl Butyrate	158.24	2.34[*]	618.7[*]	0.565[**]

[a]*Retrieved from DIPPR database [42].*
[b]*Retrieved from Ambrose et al. [43].*
[c]*Retrieved from Gude and Teja [44].*
[d]*Retrieved from Andereya and Chase [45].*
[*]*Estimated with Joback method.*
[**]*Estimated with Lee-Kessler method.*

Table 2.
Acentric factor (ω) and critical parameters (p_c,T_c) of the pure compounds used for thermodynamic modeling. From reference [30] with permission of Elsevier.

	2-Pentanol		Butyric Acid		Vinyl Butyrate		2-Pentyl Butyrate	
T (K)	k_{ij}	l_{ij}	k_{ij}	l_{ij}	k_{ij}	l_{ij}	k_{ij}	l_{ij}
Peng-Robinson - Quadratic								
313.15	0.0821	−0.0328	−0.0618	−0.1007	−0.0499	−0.0998	−0.0212	−0.0425
323.15	0.0747	−0.0371	−0.0650	−0.1178	−0.0683	−0.0706	−0.0613	−0.1196
333.15	0.0688	−0.0348	−0.0402	−0.1007	−0.0627	−0.0557	−0.0636	−0.1237
Soave-Redlich-Kwong - Quadratic								
313.15	0.0898	−0.0351	−0.0723	−0.1028	−0.0505	−0.0858	−0.0300	−0.0700
323.15	0.0873	−0.0255	−0.0743	−0.1205	−0.0631	−0.0728	−0.0501	−0.1001
333.15	0.0814	−0.0290	−0.0467	−0.1012	−0.0644	−0.0562	−0.0501	−0.1003

Table 3.
Binary interaction parameters (BIP) (k_{ij}, l_{ji}) used for the thermodynamic modeling of the (CO_2 + 2-pentanol), (CO_2 + vinyl butyrate), (CO_2 + 2-pentyl butyrate), and (CO_2 + butyric acid) binary systems at 313.15, 323.15, and 333.15 K. from reference [30] with permission of Elsevier.

	2-Pentanol	Vinyl Butyrate	2-Pentyl butyrate	Butyric Acid
T (K)		Peng-Robinson - Quadratic		
313.15	3.91	3.95	4.02	5.88
323.15	1.16	6.51	5.62	3.71
333.15	3.91	2.08	4.78	2.52
T (K)		Soave-Redlich-Kwong - Quadratic		
313.15	4.01	3.78	3.90	6.18
323.15	1.99	6.46	5.12	4.25
333.15	4.37	2.19	4.79	2.51

Table 4.
RMSD values (%) obtained for the thermodynamic modeling using Peng-Robinson and Soave-Redlich-Kwong EoS combined with quadratic mixing rules for the studied systems at 313.15, 323.15 K, and 333.15 K.

ionic liquid is used as a reaction medium to obtain (*R*)-2-pentylbutyrate. However, these conclusions have been inferred from binary mixtures, so further studies with mixtures of CO_2 with all components of the reaction would be required to corroborate the results.

4. Conclusions

Using a synthetic method in a cell of variable volume at high pressure, the phase equilibria of binary systems of (CO_2 + butyric acid), (CO_2 + 2-pentanol), (CO_2 + 2-pentyl butyrate), and (CO_2 + vinyl butyrate) at three temperatures (313.15 K, 323.15 K, and 333.15 K) have been obtained. A type I behavior was observed and three coexisting phases were not appreciated for the experimental conditions used.

A Chrastil-type correlation has been employed to correlate the $scCO_2$ solubility in organic compounds as a function of CO_2 density obtaining an equivalent accuracy to EoS-based models. The CO_2 solubility in organic compounds can be established as follows: 2-pentanol > butyric acid > esters, under the studied operating conditions. Differences in the relative polarity of the compound and CO_2 produce differences in the mutual solubility of the components.

The experimental and calculated data provided good agreement with EoS correlation of Peng-Robinson and Soave-Redlich-Kwong applied with quadratic mixing rules and two BIPs.

Author details

Mercedes G. Montalbán* and Gloria Víllora
Faculty of Chemistry, Chemical Engineering Department, University of Murcia, Regional Campus of International Excellence "Campus Mare Nostrum", Murcia, Spain

*Address all correspondence to: mercedes.garcia@um.es

References

[1] Secuianu C, Feroiu V, Geană D. Phase equilibria of carbon dioxide + 1-nonanol system at high pressures. The Journal of Supercritical Fluids. 2010;**55**(2):653-661

[2] Silva-Oliver G, Galicia-Luna LA, Sandler SI. Vapor-liquid equilibria and critical points for the carbon dioxide +1-pentanol and carbon dioxide +2-pentanol systems at temperatures from 332 to 432 K. Fluid Phase Equilibria. 2002;**200**(1):161-172

[3] Secuianu C, Feroiu V, Geană D. High-pressure vapor–liquid equilibria in the system carbon dioxide + 1-butanol at temperatures from (293.15 to 324.15) K. Journal of Chemical & Engineering Data. 2004;**49**(6):1635-1638

[4] Secuianu C, Feroiu V, Geană D. High-pressure vapor–liquid equilibria in the system carbon dioxide and 2-propanol at temperatures from 293.25 K to 323.15 K. Journal of Chemical & Engineering Data. 2003;**48**(6):1384-1386

[5] Secuianu C, Feroiu V, Geană D. High-pressure vapor–liquid and vapor–liquid–liquid equilibria in the carbon dioxide + 1-heptanol system. Fluid Phase Equilibria. 2008;**270**(1-2):109-115

[6] Fourie FC, v. N, Schwarz CE, Knoetze JH. Phase equilibria of alcohols in supercritical fluids: Part I. the effect of the position of the hydroxyl group for linear C8 alcohols in supercritical carbon dioxide. Journal of Supercritical Fluids. 2008;**47**(2):161-167

[7] Schwarz CE, Fourie FC, v. N, Knoetze JH. Phase equilibria of alcohols in supercritical fluids part II. The effect of side branching on C8 alcohols in supercritical carbon dioxide. Journal of Supercritical Fluids. 2009;**51**(2):128-135

[8] Secuianu C, Feroiu V, Geană D. Phase behavior for carbon dioxide + ethanol system: Experimental measurements and modeling with a cubic equation of state. Journal of Supercritical Fluids. 2008;**47**(2):109-116

[9] Bejarano A, Gutiérrez JE, Araus KA, de la Fuente JC. Measurement and modeling of high-pressure (vapor + liquid) equilibria of (CO2 + alkanol) binary systems. The Journal of Chemical Thermodynamics. 2011;**43**(5):759-763

[10] Gutiérrez JE, Bejarano A, de la Fuente JC. Measurement and modeling of high-pressure (vapour + liquid) equilibria of (CO2 + alcohol) binary systems. The Journal of Chemical Thermodynamics. 2010;**42**(5):591-596

[11] Nieuwoudt I, du Rand M. Measurement of phase equilibria of supercritical carbon dioxide and paraffins. Journal of Supercritical Fluids. 2002;**22**(3):185-199

[12] Schwarz CE, Knoetze JH. Phase equilibrium measurements of long chain acids in supercritical carbon dioxide. Journal of Supercritical Fluids. 2012;**66**:36-48

[13] Jang Y-S, Jeong H-H, Byun H-S. Phase equilibria for the binary mixture of n-vinyl pyrrolidone and N,N-dimethylacrylamide in supercritical carbon dioxide. Journal of Industrial and Engineering Chemistry. 2012;**18**(1):414-419

[14] Thamanavat K, Sun T, Teja AS. High-pressure phase equilibria in the carbon dioxide + pyrrole system. Fluid Phase Equilibria. 2009;**275**(1):60-63

[15] Kulik B, Kruse S, Pelto J. High pressure phase equilibrium studies

of the pyrrole–carbon dioxide binary system. Journal of Supercritical Fluids. 2008;**47**(2):135-139

[16] Pérez E, Cabañas A, Sánchez-Vicente Y, Renuncio JAR, Pando C. High-pressure phase equilibria for the binary system carbon dioxide + dibenzofuran. Journal of Supercritical Fluids. 2008;**46**(3):238-244

[17] Kim S-H, Jang Y-S, Yoon S-D, Byun H-S. High pressure phase behavior for the binary mixture of valeronitrile, capronitrile and lauronitrile in supercritical carbon dioxide at temperatures from 313.2 to 393.2K and pressures from 3.9 to 25.7MPa. Fluid Phase Equilibria. 2011;**312**:93-100

[18] Byun H-S. High pressure phase equilibria for the binary mixture of CO2+3-phenyl propionitrile and CO2+2-phenyl butyronitrile systems. Journal of Supercritical Fluids. 2017;**120**:218-225

[19] Christov M, Dohrn R. High-pressure fluid phase equilibria: Experimental methods and systems investigated (1994-1999). Fluid Phase Equilibria. 2002;**202**(1):153-218

[20] Dohrn R, Peper S, Fonseca JMS. High-pressure fluid-phase equilibria: Experimental methods and systems investigated (2000-2004). Fluid Phase Equilibria. 2010;**288**(1-2):1-54

[21] Fonseca JMS, Dohrn R, Peper S. High-pressure fluid-phase equilibria: Experimental methods and systems investigated (2005-2008). Fluid Phase Equilibria. 2011;**300**(1-2):1-69

[22] Patel RN. Enzymatic synthesis of chiral intermediates for drug development. Advanced Synthesis and Catalysis. 2001;**343**:527-546

[23] de los Ríos AP, Hernández-Fernández FJ, Tomás-Alonso F, Gómez D, Víllora G. Biocatalytic kinetic resolution of rac-1-phenylethanol and rac-2-pentanol in hexane medium: ACYL donor and water content effects. Canadian Journal of Chemical Engineering. 2010;**88**(3):442-446

[24] Lozano P, Víllora G, Gómez D, Gayo AB, Sánchez-Conesa JA, Rubio M, et al. Membrane reactor with immobilized Candida antarctica lipase B for ester synthesis in supercritical carbon dioxide. Journal of Supercritical Fluids. 2004;**29**(1-2):121-128

[25] Lozano P, De Diego T, Iborra JL, Lozano P, Diego T, Iborra JL. Biocatalytic processes using ionic liquids and supercritical carbon dioxide. In: Handbook of Green Chemistry. Weinheim, Germany: Wiley-VCH Verlag GmbH & Co. KGaA; 2010. pp. 51-73

[26] Hernández FJ, de los Ríos AP, Jiménez M, Rubio M, Gómez D, Víllora G. A new recirculating enzymatic membrane reactor for ester synthesis in ionic liquid/supercritical carbon dioxide biphasic systems. Applied Catalysis B: Environmental. 2006;**67**(1-2):121-126

[27] Lopes JM, Kareth S, Bermejo MD, Martín Á, Weidner E, Cocero MJ. Experimental determination of viscosities and densities of mixtures carbon dioxide + 1-allyl-3-methylimidazolium chloride, Viscosity correlation. The Journal of Supercritical Fluids. 2016;**111**:91-96

[28] Peng D-Y, Robinson DB. A new two-constant equation of state. Industrial and Engineering Chemistry Fundamentals. 1976;**15**(1):59-64

[29] Soave G. Equilibrium constants from a modified Redlich-Kwong equation of state. Chemical Engineering Science. 1972;**27**(6):1197-1203

[30] Carissimi G, Montalbán MG, Díaz Baños FG, Víllora G. High pressure

phase equilibria for binary mixtures of CO_2 + 2-pentanol, vinyl butyrate, 2-pentyl butyrate or butyric acid systems. Journal of Supercritical Fluids. 2018;**135**:69-77

[31] The NIST WebBook. Thermophysical properties of fluid systems. 2017. http://webbook.nist.gov/chemistry/fluid/. [Accessed November 2017]

[32] Thamanavat K. High Pressure Phase Equilibria in the Carbon Dioxide + Pyrrole System. Master thesis. USA: Georgia Institute of Technology; 2004

[33] Lee H-S, Lee H. High-pressure phase equilibria for the carbon dioxide–2-pentanol and carbon dioxide–water–2-pentanol systems. Fluid Phase Equilibria. 1998;**150-151**:695-701

[34] Byun H-S, Kim K, McHugh MA. Phase behavior and modeling of supercritical carbon dioxide–organic acid mixtures. Industrial and Engineering Chemistry Research. 2000;**39**(12):4580-4587

[35] Scott RL, van Konynenbu PH. Static properties of solutions. Van der Waals and related models for hydrocarbon mixtures. Discussions of the Faraday Society. 1970;**49**:87-97

[36] Chrastil J. Solubility of solids and liquids in supercritical gases. The Journal of Physical Chemistry. 1982;**86**(15):3016-3021

[37] Hernández EJ, Reglero G, Fornari T. Correlating the solubility of supercritical gases in high-molecular weight substances using a density-dependent equation. AICHE Journal. 2011;**57**(3):765-771

[38] Latsky C, Kouakou AC, Schwarz CE. Phase equilibria of CO2 with components in the light naphtha cut of Tyre derived oil. Journal of Supercritical Fluids. 2018;**131**:58-65

[39] Pfohl O, Petkov S, Brunner G. PE 2000: A Powerful Tool to Correlate Phase Equilibria. Germany: Herbert Utz Verlag; 2000. p. 160

[40] Nelder JA, Mead R. A simplex method for function minimization. The Computer Journal. 1965;7(4):308-313

[41] Joback KG, Reid RC. Estimation of pure-component properties from group-contributions. Chemical Engineering Communications. 1987;57(1-6):233-243

[42] DIPPR801 Database, Design Institute for Physical Properties, sponsored by AIChe. https://dippr.aiche.org/SampleDb. [Accessed November, 2017]

[43] Ambrose D, Broderick BE, Townsend R. The critical temperatures and pressures of thirty organic compounds. Journal of Applied Chemistry and Biotechnology. 2007;**24**(6):359-372

[44] Gude M, Teja AS. Vapor-liquid critical properties of elements and compounds. 4. Aliphatic alkanols. Journal of Chemical & Engineering Data. 1995;**40**(5):1025-1036

[45] Andereya E, Chase JD. The implications of carboxylic acid properties. Chemical engineering & technology - CET. 1990;**13**(1):304-312

[46] Lee BI, Kesler MG. A generalized thermodynamic correlation based on three-parameter corresponding states. AICHE Journal. 1975;**21**(3):510-527